SpringerBriefs in Computer Science

More information about this series at http://www.springer.com/series/10028

Eric Rosenberg

A Survey of Fractal Dimensions of Networks

 Springer

Eric Rosenberg
AT&T Labs
Middletown, NJ, USA

ISSN 2191-5768   ISSN 2191-5776   (electronic)
SpringerBriefs in Computer Science
ISBN 978-3-319-90046-9   ISBN 978-3-319-90047-6   (eBook)
https://doi.org/10.1007/978-3-319-90047-6

Library of Congress Control Number: 2018942025

Printed on acid-free paper

This Springer imprint is published by the registered company Springer International Publishing AG part of Springer Nature.
The registered company address is: Gewerbestrasse 11, 6330 Cham, Switzerland

To Solomon and Asher
"as an ook cometh of a litel spyr"
from "Troilus and Criseyde" (1374)
by Geoffrey Chaucer
as read by (if you listen very closely)
your great-grandfather

Preface

In the field of network science, one of the fundamental questions is how to characterize a network. One way to characterize a network is to compute a fractal dimension. We write "a fractal dimension" rather than "the fractal dimension" since many different fractal dimensions have been proposed for networks.

In this brief, we review the theory and computation of the most important of these fractal dimensions, including the box counting dimension, the correlation dimension, the mass dimension, the transfinite fractal dimension, the information dimension, the generalized dimensions (which provide a way to describe multifractal networks), and the sandbox method (for approximating the generalized dimensions). We describe the use of diameter-based and radius-based boxes, and present several heuristic methods for box counting, including greedy coloring, random sequential node burning, and a method for computing a lower bound. We also discuss very recent results on resolving ambiguity in the calculation of the information dimension and the generalized dimensions, and on the non-monotonicity of the generalized dimensions.

Research on fractal dimensions of networks began at least as early as 1988. Since about 2003, this has been an active research area. The wide variety of fractal dimensions considered in this survey is a testament to the richness of this subject. There are abundant opportunities for research and applications.

The intended audience for this book is anyone interested in the theory and application of networks. This includes anyone studying, e.g., social networks, telecommunications networks, transportation networks, ecological networks, food chain networks, network models of the brain, or financial networks. We assume a knowledge of limits and the derivative of a function of a single variable. We also assume the reader is familiar with finding the shortest path between two nodes in a network, e.g., using Dijkstra's method. A knowledge of the theory of linear programming, specifically, duality theory and complementary slackness, is required for Chap. 4.

Many thanks to Kartik Pandit and Curtis Provost, and a special acknowledgment to Robert Murray, for their comments and suggestions on this survey. Many thanks also to Paul Drougas, Senior Editor at Springer, and to Frank Politano, Esq of K&L Gates LLP, for bringing this brief to fruition.

Middletown, NJ, USA   Eric Rosenberg
March 2018

About the Author

Eric Rosenberg received a B.A. in Mathematics from Oberlin College and a Ph.D. in Operations Research from Stanford University. He works at AT&T Labs in Middletown, New Jersey (email: ericr@att.com). Dr. Rosenberg has taught undergraduate and graduate courses in optimization at Princeton University and New Jersey Institute of Technology. He has authored or coauthored 17 patents and has published in the areas of convex analysis and nonlinearly constrained optimization, computer-aided design of integrated circuits and printed wire boards, telecommunications network design and routing, and fractal dimensions of networks.

Contents

1 Introduction 1
  1.1 Tables of Symbols 2
  1.2 Box Counting and Hausdorff Dimensions 2

2 Covering a Complex Network 7
  2.1 Box Counting with Diameter-Based or Radius-Based Boxes 9

3 Network Box Counting Heuristics 13
  3.1 Node Coloring Formulation 13
  3.2 Node Coloring for Weighted Networks 15
  3.3 Random Sequential Node Burning 17
  3.4 Set Covering Formulation and a Greedy Method 19
  3.5 Box Burning 22
  3.6 Box Counting for Scale-Free Networks 25

4 Lower Bounds on Box Counting 29
  4.1 Mathematical Formulation 29
  4.2 Dual Ascent and Dual Adjustment 32
  4.3 Bounding the Fractal Dimension 36

5 Correlation Dimension 39

6 Mass Dimension for Infinite Networks 45
  6.1 Transfinite Fractal Dimension 48

7 Volume and Surface Dimensions for Infinite Networks 51

8 Information Dimension 55

9 Generalized Dimensions 61

10 Non-monotonicity of Generalized Dimensions 69

11 Zeta Dimension 77

References 81

Chapter 1
Introduction

Consider the network $\mathbb{G} = (\mathbb{N}, \mathbb{A})$ where $\mathbb{N}$ is the set of nodes connected by the set $\mathbb{A}$ of arcs. Let $N \equiv |\mathbb{N}|$ be the number of nodes, and let $A \equiv |\mathbb{A}|$ be the number of arcs. (We use "$\equiv$" to denote a definition.) Unless otherwise specified, henceforth we assume that $\mathbb{G}$ is a "complex network": an arbitrary network without special structure (as opposed to, e.g., a regular lattice), for which all arcs have unit cost (so the length of a shortest path between two nodes is the number of arcs in that path), and all arcs are undirected (so the arc between nodes i and j can be traversed in either direction). We assume that $\mathbb{G}$ is *connected*, meaning there is a path of arcs in $\mathbb{A}$ connecting any two nodes in $\mathbb{N}$. Many different measures have been used to describe a complex network [7]. For example,

1. the *density* of $\mathbb{G}$ is $A/[N(N-1)/2]$, which is the ratio of the number of arcs in $\mathbb{G}$ to the number of arcs that would be present if each pair of nodes were connected by an arc;
2. the *diameter* Δ of $\mathbb{G}$ is defined by $\Delta \equiv \max\{\, dist(x, y) \,|\, x, y \in \mathbb{N}\}$, where $dist(x, y)$ is the length of the shortest path between nodes x and y;
3. the *average node degree* is $(1/N) \sum_{n \in \mathbb{N}} \delta_n$, where δ_n is the node degree of node n (the number of arcs incident to node n).

Another way to describe $\mathbb{G}$ is to compute a *fractal dimension* of $\mathbb{G}$. The interest in computing fractal dimensions of networks began about 2003 [10] (although one important paper [38] appeared as early as 1988) and was inspired by the study of fractal dimensions of geometric objects. For geometric objects, the fractal dimensions of interest usually have non-integer values, and for that reason a fractal is sometimes defined as something with a non-integer dimension. However, a good definition of "a fractal" has been an elusive goal, and research has instead focused on the theory and application of different fractal dimensions. Similarly, we are not concerned here with a definition of "a fractal network". Rather, our goal is to survey the most important of the different fractal dimensions that have been proposed for networks.

© The Author(s), under exclusive licence to Springer International Publishing AG, part of Springer Nature 2018
E. Rosenberg, *A Survey of Fractal Dimensions of Networks*, SpringerBriefs in Computer Science, https://doi.org/10.1007/978-3-319-90047-6_1

This book is organized as follows. In this chapter we will present the box counting and Hausdorff dimensions. Chapter 2 considers the covering of a complex network by boxes. Several heuristics for box counting are presented in Chap. 3, and Chap. 4 is concerned with lower bounds on box counting. The correlation dimension is presented in Chap. 5. The mass dimension and the transfinite fractal dimension of an infinite network are studied in Chap. 6, and the volume and surface dimensions of an infinite network are studied in Chap. 7. The information dimension is the subject of Chap. 8. Generalized dimensions and the sandbox method are studied in Chap. 9. The non-monotonicity of the generalized dimensions is explored in Chap. 10. The final chapter, Chap. 11, presents the zeta dimension.

1.1   Tables of Symbols

Many of the fractal dimensions studied in this book are defined in terms of a covering of $\mathbb{G}$ by "boxes", and the reason for covering $\mathbb{G}$ by boxes is that this is how the fractal dimension of a geometric object was first defined. So we start, in the next section, with a review of the box counting and Hausdorff dimensions of a geometric object. For convenience, some symbols frequently used in this survey are summarized in Table 1.1, and a list of the fractal dimensions we study is provided in Table 1.2. The end of a definition, or proof, or example, is indicated by the symbol □.

1.2   Box Counting and Hausdorff Dimensions

The simplest fractal dimension is the box counting dimension, which is based on covering a geometric object $\Omega \subset \mathbb{R}^E$ by equal sized E-dimensional hypercubes. For example, for $E = 1$, consider a line segment of length L. If we measure the line segment using a ruler of length s, where $s \ll L$, the number $B(s)$ of rule lengths needed is given by $B(s) \approx Ls^{-1}$. We call $B(s)$ the "number of boxes" of size s needed to cover the segment. For $E = 1$, a "box" of size s is a line segment of length s. Since the exponent of s in Ls^{-1} is -1, we say that a line segment has a *box counting* dimension of 1. Now consider a two-dimensional square with side length L. If we cover the square by small squares of side length s, where $s \ll L$, the number $B(s)$ of small squares needed is given by $B(s) \approx L^2 s^{-2}$. Since the exponent of s in $L^2 s^{-2}$ is -2, we say that the square has box counting dimension 2.

To provide a general definition of the box counting dimension of a geometric object, let Ω be a closed and bounded subset of $\mathbb{R}^E$. By the "linear size" of Ω we mean the diameter of Ω (the maximal Euclidean distance between any two points of Ω, denoted by $diam(\Omega)$), or the maximal variation in any coordinate (i.e., $\max_{x,y \in \Omega} \max_{1 \leq i \leq E} |x_i - y_i|$). Let $s \ll 1$ be the linear size of a small box, where a "box" is an E-dimensional hypercube. By a box of size s we mean a box of linear size s. A set of boxes *covers* Ω if each point in Ω belongs to at least one box.

Table 1.1 Symbols and their definitions

Symbol	Definition
δ_n	Node degree of node n
Δ	Network diameter
Ω	Geometric object in $\mathbb{R}^E$
$\mathbb{A}$	Set of arcs in $\mathbb{G}$
$\mathscr{B}(s)$	A minimal s-covering of $\mathbb{G}$
$B(s)$	Cardinality of $\mathscr{B}(s)$
$B_D(s)$	Minimal number of diameter-based boxes needed to cover $\mathbb{G}$
$B_R(r)$	Minimal number of radius-based boxes needed to cover $\mathbb{G}$
B_j	Box in $\mathscr{B}(s)$
$\mathbb{G}$	Complex network
$\mathbb{G}(n, r)$	Subnetwork of $\mathbb{G}$ with center n and radius r
$\mathbb{G}(s)$	Subnetwork of $\mathbb{G}$ of diameter $s-1$
$H(s)$	Entropy of the probability distribution $p_j(s)$
$M(n, r)$	Number of nodes in $\mathbb{N}(n, r)$
$\mathbb{N}$	Set of nodes in $\mathbb{G}$
N	Number of nodes in $\mathbb{G}$
$N_j(s)$	Number of nodes in box $B_j \in \mathscr{B}(s)$
$\mathbb{N}(n, r)$	The set of nodes whose distance from n does not exceed r
$p_j(s)$	Probability of box $B_j \in \mathscr{B}(s)$
$\mathbb{R}^E$	E-dimensional Euclidean space
$x(s)$	Vector summarizing the covering $\mathscr{B}(s)$
$Z_q\big(\mathscr{B}(s)\big)$	Partition function value for the covering $\mathscr{B}(s)$
$Z(x, q)$	Partition function value for the summary vector x

Table 1.2 Fractal dimensions

Symbol	Definition
d_B	Box counting dimension
d_C	Correlation dimension
d_E	Transfinite fractal dimension
d_I	Information dimension
d_M	Mass dimension
d_U	Surface dimension
d_V	Volume dimension
d_Z	Zeta dimension
D_q	Generalized dimension of order q
$D_q^{sandbox}$	Sandbox dimension of order q
$D_q(L, U)$	Secant estimate of D_q

Definition 1.1 Let $B(s)$ be the minimal number of boxes of size s needed to cover Ω. If

$$\lim_{s \to 0} \frac{\log B(s)}{\log(1/s)} \tag{1.1}$$

exists, then the limit is called the *box counting dimension* of Ω and is denoted by d_B. $\square$

Roughly speaking, if d_B is the box counting dimension of Ω then $B(s)$ behaves as s^{-d_B} for $s \ll 1$. In practice, the computation of d_B typically begins by selecting a set $\{s_1, s_2, \cdots, s_K\}$ of box sizes. For each value of s, we determine the minimal number $B(s)$ of boxes of size s needed to cover Ω. By plotting $\log B(s)$ vs. $\log s$ for the K values of s, a range of s can be identified over which the plot is roughly linear [3, 27, 28, 45]. Then d_B can be determined, e.g., by linear regression.

Although the limit (1.1) may not exist [12], the lim inf and lim sup always exist. The *lower box counting dimension* $\underline{d_B}$ is defined by

$$\underline{d_B} \equiv \liminf_{s \to 0} \frac{\log B(s)}{\log(1/s)}, \tag{1.2}$$

and the *upper box counting dimension* $\overline{d_B}$ is defined by

$$\overline{d_B} \equiv \limsup_{s \to 0} \frac{\log B(s)}{\log(1/s)}. \tag{1.3}$$

When $\underline{d_B} = \overline{d_B}$ then d_B exists, and $d_B = \underline{d_B} = \overline{d_B}$.

The box counting dimension assumes that all the boxes used to cover Ω are identical. This restriction is removed in the *Hausdorff* dimension [14, 53, 63], introduced in 1918 by Felix Hausdorff (1868–1942) [24]. Due to the contributions of Abram Samoilovitch Besicovitch (1891–1970), this dimension is sometimes called the *Hausdorff-Besicovitch* dimension. For $s > 0$, define an s-covering of Ω to be a finite collection of J sets $\{X_1, X_2, \cdots, X_J\}$ that cover Ω (i.e., $\Omega \subseteq \cup_{j=1}^{J} X_j$) such that for each j we have $diam(X_j) \leq s$. Let $\mathscr{C}(s)$ be the set of all s-coverings of Ω. For $d > 0$, define

$$v(d, s) \equiv \inf_{\mathscr{C}(s)} \sum_{j=1}^{J} \left(diam(X_j)\right)^d, \tag{1.4}$$

where the infimum is over all s-coverings $\mathscr{C}(s)$ of Ω. We take the infimum since the goal is to cover Ω with small sets X_j as efficiently as possible.

We can think of $v(d, s)$ as the d-dimensional volume of Ω. For almost all values of d, the limit $\lim_{s \to 0} v(d, s)$ is either 0 or ∞, where by ∞ we mean $+\infty$. For example, suppose we cover the unit square $[0, 1] \times [0, 1]$ by small squares of side length s. We need $1/s^2$ small squares, the diameter of each square is $\sqrt{2}s$, and $v(d, s) = (1/s^2)(\sqrt{2}s)^d = \sqrt{2}^d s^{d-2}$. We have

$$\lim_{s \to 0} s^{d-2} = \begin{cases} \infty & \text{if } d < 2 \\ 1 & \text{if } d = 2 \\ 0 & \text{if } d > 2. \end{cases}$$

Thus, for example, if $d = 3$ then the unit square $[0, 1] \times [0, 1]$ has zero volume; if $d = 1$ then the unit square has infinite length.

For a given d, as s decreases, the set of available covers shrinks, so $v(d, s)$ increases as s decreases. Thus

$$v^\star(d) \equiv \lim_{s \to 0} v(d, s) \tag{1.5}$$

always exists in $[0, \infty) \cup \{\infty\}$; that is, $v^\star(d)$ might be ∞. (We call $v^\star(d)$ the d-dimensional Hausdorff measure of Ω.) Since for each fixed $s < 1$ the function $v(d, s)$ is non-increasing with d, then $v^\star(d)$ is also non-increasing with d. For $d \geq 0$ and $d' \geq 0$, definition (1.5) implies [53]

If $v^\star(d) < \infty$ and $d' > d$, then $v^\star(d') = 0$.

If $v^\star(d) > 0$ and $d' < d$, then $v^\star(d') = \infty$.

These two assertions imply the existence of a unique value of d, called the *Hausdorff dimension* of Ω and denoted by d_H, such that $v^\star(d) = \infty$ for $d < d_H$ and $v^\star(d) = 0$ for $d > d_H$. Formally,

$$d_H \equiv \inf\{d \geq 0 \,|\, v^\star(d) = 0\}. \tag{1.6}$$

The Hausdorff dimension d_H of Ω might be zero, positive, or ∞.

Mandelbrot [36, p. 15] uses the term "fractal dimension" to refer to the Hausdorff dimension; we will use the term "fractal dimension" only in a generic manner, to refer to any fractal dimension. Although historically the Hausdorff dimension preceded the box counting dimension, the Hausdorff dimension is a generalization of the box counting dimension, since the Hausdorff dimension does not require equal size boxes, while the box counting dimension does. A set $\Omega \subset \mathbb{R}^E$ with $d_H < 1$ is totally disconnected [12] (a set Ω is totally disconnected if for $x \in \Omega$, the largest connected component of Ω containing x is x itself). If $\Omega \subset \mathbb{R}^E$ is an open set, then $d_H = E$. If Ω is countable, then $d_H = 0$. For ordinary geometric objects, the Hausdorff and box counting dimensions are equal: $d_H = d_B$. However, in general, $d_H \leq d_B$ [12]. This inequality holds since the set $\mathscr{C}(s)$ of all s-coverings of Ω includes any covering of Ω by boxes of size s.

The above discussion suggests that it is easier to obtain an upper bound on d_H than a lower bound. Obtaining a lower bound on d_H requires estimating all possible coverings of Ω. Also, the box counting dimension is not as well-behaved as the Hausdorff dimension. For example, a countable set can have a positive d_B [12]. However, in practice the Hausdorff dimension has been rarely used, while the box counting dimension has been widely used, since it is easier to compute.

Chapter 2
Covering a Complex Network

The previous chapter showed how the box counting and Hausdorff dimensions of a geometric object Ω are computed from a covering of Ω. With this background, we can now consider what it means to cover a complex network $\mathbb{G}$, and how a fractal dimension can be computed from a covering of $\mathbb{G}$. We require some definitions. The network B is a *subnetwork* of $\mathbb{G}$ if B can be obtained from $\mathbb{G}$ by deleting nodes and arcs. By a *box* we mean a subnetwork of $\mathbb{G}$. A box is *disconnected* if some nodes in the box cannot be connected by arcs in the box. Let $\{B_j\}_{j=1}^{J} \equiv \{B_1, B_2, \cdots, B_J\}$ be a collection of boxes. Two types of coverings of $\mathbb{G}$ have been proposed: *node coverings* and *arc coverings*. Let s be a positive integer.

Definition 2.1 (*i*) The set $\{B_j\}_{j=1}^{J}$ is a *node s-covering* of $\mathbb{G}$ if for each j we have $diam(B_j) < s$ and if each node in $\mathbb{N}$ is contained in exactly one B_j. (*ii*) The set $\{B_j\}_{j=1}^{J}$ is an *arc s-covering* of $\mathbb{G}$ if for each j we have $diam(B_j) < s$ and if each arc in $\mathbb{A}$ is contained in exactly one B_j. $\square$

If B_j is a box in a node or arc s-covering of $\mathbb{G}$ then the requirement $diam(B_j) < s$ in Definition 2.1 implies that B_j is connected. However, this requirement, which is a standard assumption in defining the box counting dimension of $\mathbb{G}$ (e.g., [16, 29, 30, 48, 56]), may frequently be violated, for good reasons, in some methods for determining the fractal dimensions of $\mathbb{G}$, as we will discuss in Sect. 3.6.

It is possible to define a node covering of $\mathbb{G}$ to allow a node to be contained in more than one box; coverings with possibly overlapping boxes are studied in [15, 60]. The great advantage of non-overlapping boxes is that they immediately yield a probability distribution, as discussed in Chap. 8. The probability distribution obtained from a non-overlapping node covering of $\mathbb{G}$ is the basis for computing the information dimension d_I and the generalized dimensions D_q of $\mathbb{G}$ (Chap. 9). Therefore, in this survey, each node covering of $\mathbb{G}$ is assumed to use non-overlapping boxes, as specified in Definition 2.1.

© The Author(s), under exclusive licence to Springer International Publishing AG, part of Springer Nature 2018
E. Rosenberg, *A Survey of Fractal Dimensions of Networks*, SpringerBriefs in Computer Science, https://doi.org/10.1007/978-3-319-90047-6_2

Fig. 2.1 A node 3-covering (**a**) and an arc 3-covering (**b**)

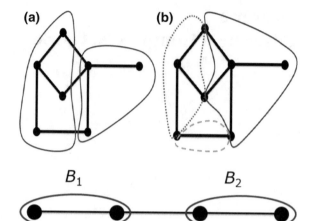

Fig. 2.2 A node covering which is not an arc covering

B_1 B_2

Since a network of diameter 0 contains only a single node, a node 1-covering contains N boxes. Since a node 1-covering provides no useful information other than N itself, we consider node s-coverings only for $s \geq 2$. Figure 2.1a illustrates a node 3-covering with $J = 2$. Both boxes in the 3-covering have diameter 2. Figure 2.1b illustrates an arc 3-covering of the same network using three boxes. The box indicated by the solid blue line contains four arcs, the box indicated by the dotted red line contains three arcs, and the box indicated by the dashed green line contains one arc.

Each arc covering of $\mathbb{G}$ yields a node covering. However, the converse is not true: a node covering does not in general yield an arc covering. This is illustrated by the simple example of Fig. 2.2. The nodes are covered by B_1 and B_2, but the arc in the middle belongs to neither B_1 nor B_2.

Definition 2.2 (*i*) An arc s-covering $\{B_j\}_{j=1}^{J}$ is *minimal* if for any other arc s-covering $\{B_j'\}_{j=1}^{J'}$ we have $J \leq J'$. (*ii*) A node s-covering $\{B_j\}_{j=1}^{J}$ is *minimal* if for any other node s-covering $\{B_j'\}_{j=1}^{J'}$ we have $J \leq J'$. $\square$

That is, a covering is minimal if it uses the fewest possible number of boxes. For $s > \Delta$, the minimal node or arc s-covering consists of a single box, which is $\mathbb{G}$ itself. Virtually all research has considered node coverings; only a few studies (e.g., [74]) use arc coverings. The reason arc coverings are rarely used is that, in practice, computing a fractal dimension of a geometric object typically starts with a given set of points in $\mathbb{R}^E$ (the points are then covered by boxes, or the distance between each pair of points is computed (e.g., [22, 36])), and nodes in a network are analogous to points in $\mathbb{R}^E$. Having very briefly contrasted arc coverings and node coverings for a network, we now abandon arc coverings; henceforth, all coverings of $\mathbb{G}$ are node coverings, and by covering $\mathbb{G}$ we mean covering the nodes of $\mathbb{G}$. Also, henceforth by an s-covering we mean a node s-covering, and by a covering of size s we mean an s-covering.

2.1   Box Counting with Diameter-Based or Radius-Based Boxes

There are two main approaches used to define boxes for use in covering $\mathbb{G}$: diameter-based boxes and radius-based boxes.

Definition 2.3 (*i*) A radius-based box $\mathbb{G}(n, r)$ with center node $n \in \mathbb{N}$ and radius r is the subnetwork of $\mathbb{G}$ containing all nodes whose distance to n does not exceed r. Let $B_R(r)$ be the minimal number of radius-based boxes of radius at most r needed to cover $\mathbb{G}$. (*ii*) A diameter-based box $\mathbb{G}(s)$ of size s is a subnetwork of $\mathbb{G}$ of diameter $s - 1$. Let $B_D(s)$ denote the minimal number of diameter-based boxes of size at most s needed to cover $\mathbb{G}$.   □

Thus the node set of $\mathbb{G}(n, r)$ is $\{x \in \mathbb{N} \,|\, dist(n, x) \le r\}$. Radius-based boxes are used in the *Maximum Excluded Mass Burning* and *Random Sequential Node Burning* methods described in Chap. 3. Interestingly, the above definition of a radius-based box may frequently be violated in the *Maximum Excluded Mass Burning* and *Random Sequential Node Burning* methods. In particular, some radius-based boxes created by those methods may be disconnected, or some boxes may contain only some of the nodes whose distance to the center node n does not exceed r.

A diameter-based box $\mathbb{G}(s)$ is not defined in terms of a center node; instead, for $x, y \in \mathbb{G}(s)$ we require $dist(x, y) < s$. Diameter-based boxes are used in the *Box Burning* and *Compact Box Burning* heuristics described in Chap. 3. The above definition of a diameter-based box also may frequently be violated in the *Box Burning* and *Compact Box Burning* methods. Also, since each node in $\mathbb{G}$ must belong to exactly one B_j in an s-covering $\{B_j\}_{j=1}^{J}$ using diameter-based boxes, then in general we will not have $diam(B_j) = s - 1$ for all j. To see this, consider a chain of three nodes (call them x, y, and z), and let $s = 2$. The minimal 2-covering using diameter-based boxes requires two boxes, B_1 and B_2. If B_1 covers x and y then B_2 covers only z, so the diameter of B_2 is 0.

The minimal number of diameter-based boxes of size at most $2r + 1$ needed to cover $\mathbb{G}$ is, by definition, $B_D(2r + 1)$. We have $B_D(2r + 1) \le B_R(r)$ [29]. To see this, let $\mathbb{G}(n_j, r_j)$, $j = 1, 2, \cdots, B_R(r)$ be the boxes in a minimal covering of $\mathbb{G}$ using radius-based boxes of radius at most r. Then $r_j \le r$ for all j. Pick any j, and consider box $\mathbb{G}(n_j, r_j)$. For any nodes x and y in $\mathbb{G}(n_j, r_j)$ we have

$$dist(x, y) \le dist(x, n_j) + dist(n_j, y) \le 2r_j \le 2r ,$$

so $\mathbb{G}(n_j, r_j)$ has diameter at most $2r$. Thus these $B_R(r)$ boxes also serve as a covering of size $2r + 1$ using diameter-based boxes. Therefore, the minimal number of diameter-based boxes of size at most $2r + 1$ needed to cover $\mathbb{G}$ cannot exceed $B_R(r)$; that is, $B_D(2r + 1) \le B_R(r)$.

Fig. 2.3 Diameter-based vs.
radius-based boxes

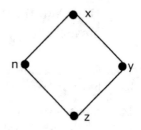

The reverse inequality does not in general hold, since a diameter-based box of size $2r+1$ can contain more nodes than a radius-based box of radius r. For example, consider the network $\mathbb{G}$ of Fig. 2.3. The only nodes adjacent to n are x and z, so $\mathbb{G}(n, 1) = \{n, x, z\}$ and $B_R(1) = 2$. Yet the diameter of $\mathbb{G}$ is 2, so it can be covered by a single diameter-based box of size 3, namely $\mathbb{G}$ itself, so $B_D(3) = 1$. Thus $B_R(r)$ and $B_D(2r+1)$ are not in general equal. Nonetheless, for the *C. elegans* and Internet backbone networks studied in [56], the calculated fractal dimension was the same whether radius-based or diameter-based boxes were used. Similarly, both radius-based and diameter-based boxes yielded a fractal dimension of approximately 4.1 for the WWW (the World Wide Web) [29].

The term *box counting* refers to computing a minimal s-covering of $\mathbb{G}$ for a range of values of s, using either radius-based boxes or diameter-based boxes. Conceivably, other types of boxes might be used to cover $\mathbb{G}$. In the fractal literature, the box counting dimension d_B is often informally defined by the scaling $B_D(s) \sim s^{-d_B}$. (The symbol "$\sim$", frequently used in the fractal literature but often with different meanings, should here be interpreted to mean "approximately behaves like".) Definition 2.4 below provides a more computationally useful definition of d_B for a complex network.

Definition 2.4 $\mathbb{G}$ has box counting dimension d_B if over some range of s and for some constant c we have

$$\log B_D(s) \approx -d_B \log(s/\Delta) + c. \quad \square \tag{2.1}$$

Alternatively, (2.1) can be written as $\log B_D(s) \approx -d_B \log s + c$. If $\mathbb{G}$ has box counting dimension d_B then over some range of s we have $B_D(s) \approx as^{-d_B}$ for some constant a. In the terminology of [16], if the box counting dimension for $\mathbb{G}$ exists, then $\mathbb{G}$ enjoys the *fractal scaling property*, or, more simply, $\mathbb{G}$ is *fractal*. The main feature apparently displayed by fractal networks is a repulsion between hubs, where a hub is a node with a significantly higher node degree than a non-hub node. That is, the highly connected nodes tend to be not directly connected [72]. This tendency can be quantified using the joint node degree distribution $p(\delta_1, \delta_2)$ that a node with degree δ_1 and a node with degree δ_2 are neighbors (i.e., connected by a single arc). In contrast, for a non-fractal network $\mathbb{G}$, hubs are mostly connected to other hubs, which implies that $\mathbb{G}$ enjoys the small-world property [16]. (Roughly speaking, $\mathbb{G}$ is a *small-world* network if $diam(G)$ grows as $\log(N)$ [57].) Also, the concepts of

Fig. 2.4 Fractal vs.
non-fractal scaling

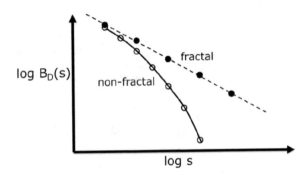

modularity and *fractality* for a network are closely related. Interconnections within a module (e.g., a biological sub-system) are more prevalent than interconnections between modules. Similarly, in a fractal network, interconnections between a hub and non-hub nodes are more prevalent than interconnections between hubs. Non-fractal networks are typically characterized by a sharp decay of $B_D(s)$ with s, which is better described by an exponential law $B_D(s) \sim e^{-\beta s}$, where $\beta > 0$, rather than by a power law $B_D(s) \sim s^{-\beta}$, with a similar statement holding if radius-based boxes are used. These two cases are illustrated in Fig. 2.4, taken from [16], where the solid circles are measurements from a fractal network, and the hollow circles are from a non-fractal network.

Chapter 3
Network Box Counting Heuristics

In this chapter we examine several methods for computing a minimal set of diameter-based boxes, or a minimal set of radius-based boxes, to cover $\mathbb{G}$. We will see that some of these methods, which have been shown to be quite effective, nonetheless require us to bend some of the definitions presented in Chap. 2. In particular, some of these methods may generate boxes that are not connected subnetworks of $\mathbb{G}$.

3.1 Node Coloring Formulation

The problem of determining the minimal number $B_D(s)$ of diameter-based boxes of size at most s needed to cover $\mathbb{G}$ is an example of the NP-hard *graph coloring problem* [56], for which many good heuristics are available. To transform the covering problem into the graph coloring problem for a given $s \geq 2$, first create the auxiliary graph $\widetilde{\mathbb{G}}_s = (\mathbb{N}, \widetilde{\mathbb{A}}_s)$ as follows. The node set of $\widetilde{\mathbb{G}}_s$ is $\mathbb{N}$; it is independent of s. The arc set $\widetilde{\mathbb{A}}_s$ of $\widetilde{\mathbb{G}}_s$ depends on s: there is an undirected arc (u, v) in $\widetilde{\mathbb{A}}_s$ if $dist(u, v) \geq s$, where the distance is in the original graph $\mathbb{G}$.

Having constructed $\widetilde{\mathbb{G}}_s$, the task is to color the nodes of $\widetilde{\mathbb{G}}_s$, using the minimal number of colors, such that no arc in $\widetilde{\mathbb{A}}_s$ connects nodes assigned the same color. That is, if $(u, v) \in \widetilde{\mathbb{A}}_s$, then u and v must be assigned different colors. The minimal number of colors required is called the *chromatic number* of $\widetilde{\mathbb{G}}_s$, traditionally denoted by $\chi(\widetilde{\mathbb{G}}_s)$.

Theorem 3.1 $\chi(\widetilde{\mathbb{G}}_s) = B_D(s)$.

Proof ([56]) Suppose that nodes u and v are assigned the same color. Then u and v cannot be the endpoints of an arc in $\widetilde{\mathbb{G}}_s$, because if they were, they would be assigned different colors. Hence $dist(u, v) < s$, so u and v can be placed in a single

© The Author(s), under exclusive licence to Springer International Publishing AG, part of Springer Nature 2018
E. Rosenberg, *A Survey of Fractal Dimensions of Networks*, SpringerBriefs in Computer Science, https://doi.org/10.1007/978-3-319-90047-6_3

Fig. 3.1 Example network
for node coloring

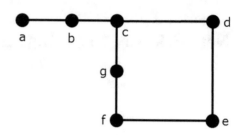

diameter-based box of size s. It follows that $B_D(s) \leq \chi(\widetilde{\mathbb{G}}_s)$. To prove the reverse
inequality, consider any minimal s-covering using $B_D(s)$ boxes, and let B be any
box in this covering. For any nodes x and y in this box we have $dist(x, y) < s$, so
x and y are not connected by an arc in $\widetilde{\mathbb{G}}_s$. Thus x and y can be assigned the same
color, which implies $\chi(\widetilde{\mathbb{G}}_s) \leq B_D(s)$. $\quad\square$

We illustrate the node coloring formulation using the network of Fig. 3.1. For $s =$
3, the auxiliary graph $\widetilde{\mathbb{G}}_3$ is given by Fig. 3.2a. Arc (x, y) exists in $\widetilde{\mathbb{G}}_3$ if $dist(x, y) \geq$
3. Thus node c is isolated in $\widetilde{\mathbb{G}}_3$ since its distance in $\mathbb{G}$ to all other nodes does not
exceed 2. Also, the distance in $\mathbb{G}$ from g to all nodes except a does not exceed 2, so
arc (g, a) exists in $\widetilde{\mathbb{G}}_3$.

The chromatic number $\chi(\widetilde{\mathbb{G}}_3)$ of the simple network of Fig. 3.2a can be exactly
computed using the *Greedy Coloring* method [56], which assigns colors based on a
random ordering of the nodes. Typically *Greedy Coloring* would be run many times,
using different random orderings of the nodes; using 10,000 random orderings,
Greedy Coloring has been shown to provide significant accuracy. Moreover the
method is very efficient, since, for a given random ordering of the nodes, a single
pass through all the nodes suffices to compute an s-covering of $\mathbb{G}$ for all box sizes
s [56].

We illustrate *Greedy Coloring* using Fig. 3.2b. Suppose we randomly pick a as
the first node, and assign the color yellow to node a (a yellow node is indicated
using a small *square* box). Then d, e, f, and g cannot be colored yellow, so we
color them blue (a blue node is indicated using a small *oval* box). We can color
b yellow since it is connected only to nodes already colored blue. Since c is
isolated we are free to assign it any color, so we color it yellow. We are done;
nodes a, b, and c are in the yellow box and nodes d, e, f, and g are in the blue
box. This is an optimal coloring, since at least two colors are needed to color any
graph with at least one arc. Figure 3.3 illustrates, in the original network, the two
boxes in this minimal covering for $s = 3$.

For $s = 4$, the auxiliary graph $\widetilde{\mathbb{G}}_4$ is shown in Fig. 3.4a. There is an arc (x, y)
in $\widetilde{\mathbb{G}}_4$ if in $\mathbb{G}$ we have $dist(x, y) \geq 4$. We again apply the *Greedy Coloring*
heuristic to compute the chromatic number $\chi(\widetilde{\mathbb{G}}_4)$. Suppose we randomly pick a
as the first node, and assign the color yellow to node a. Then e and f cannot
be colored yellow, so we color them blue. The remaining nodes are isolated
so we arbitrarily color them blue. We are done; a is in the yellow box and the
remaining nodes are in the blue box. This is also an optimal coloring. Figure 3.4b
illustrates, in the original network, the two boxes in the minimal covering for $s = 4$.

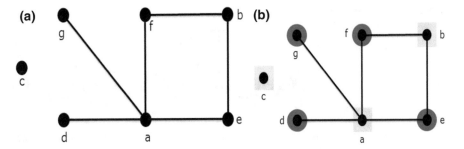

Fig. 3.2 (**a**) Auxiliary graph with $s=3$, and (**b**) its coloring

Fig. 3.3 Minimal covering for $s = 3$

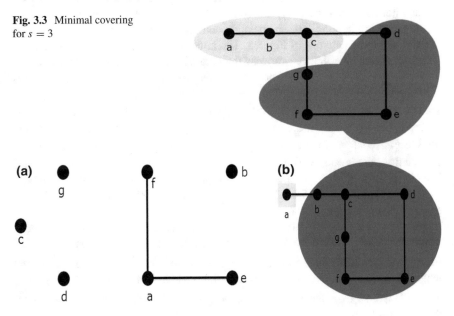

Fig. 3.4 (**a**) Auxiliary graph with $s=4$, and (**b**) its coloring

3.2 Node Coloring for Weighted Networks

For an unweighted network, the distance between two nodes (also known as the chemical distance, or the hop count) ranges from 1 to the diameter Δ of the network. However, when applying box counting to a weighted network, choosing box sizes between 2 and Δ will not in general be useful. For example, if the network diameter is less than 1, then the entire network is contained in a box of size 1. One simple approach to dealing with box size selection for weighted networks is to multiply each arc length by a sufficiently large constant k. For example, if we approximate each arc length by a rational number, then choosing k to be the least common denominator of all these rational numbers will yield a set of integer arc lengths.

Fig. 3.5 A weighted network

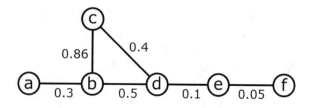

Even if all the arc lengths are integer, a set of box sizes must still be selected. The box sizes could be selected using a default method, such as starting with a box size less than the network diameter, and decreasing the box size by a factor of 2 in each iteration. Alternatively, the box sizes could be determined by an analysis of the set of arc lengths. This is the approach taken in [69], and we describe their method using the network of Fig. 3.5. The nodes are a, b, c, d, e, f, and each arc length is shown. We first pre-process the data by computing the shortest distance d_{ij} between each pair (i, j) of nodes. The largest d_{ij} is the diameter Δ. For this example we have $\Delta = 1.16$, which is the distance between nodes a and c. The second pre-processing step is to sort all the d_{ij} values in increasing order. For this example, the five smallest d_{ij} values are $0.05, 0.1, 0.3, 0.4, 0.5$. Next we compute the successive sums of the ordered d_{ij} values, stopping when the sum first exceeds Δ. The first sum is $\sigma(1) = 0.05$, the second is $\sigma(2) = \sigma(1) + 0.1 = 0.15$, the third is $\sigma(3) = \sigma(2) + .3 = .45$, the fourth is $\sigma(4) = \sigma(3) + 0.4 = 0.85$, and finally $\sigma(5) = \sigma(4) + 0.5 = 1.35 > \Delta$. We set $s_1 = \sigma(4) = 0.85$ since this is the largest sum not exceeding Δ.

Next we create an auxiliary graph $\widetilde{\mathbb{G}}$ such that an arc in $\widetilde{\mathbb{G}}$ exists between nodes i and j if $d_{ij} \geq s_1$. There are four pairs of nodes for which $d_{ij} \geq s_1$, namely (a, c), (a, e), (a, f), and (c, b), so $\widetilde{\mathbb{G}}$ has four arcs. Node d does not appear in $\widetilde{\mathbb{G}}$. The length of arc (i, j) in $\widetilde{\mathbb{G}}$ is d_{ij}, e.g., $d_{af} = 0.95$, which is the length of the shortest path in $\mathbb{G}$ from a to f. Next we assign a weight to each node in $\widetilde{\mathbb{G}}$. For node i in $\widetilde{\mathbb{G}}$, the weight $w(i)$ is

$$ w(i) \equiv \sum_{(i,j)\in\widetilde{\mathbb{G}}} d_{ij} . $$

The $w(i)$ values are the underlined values in Fig. 3.6 next to each node. Thus $w(a) = 0.9 + 1.16 + 0.95 = 3.01$ for the three arcs in $\widetilde{\mathbb{G}}$ incident to a, and $w(f) = 0.95$ for the one arc in $\widetilde{\mathbb{G}}$ incident to f. As in Sect. 3.1, we color the nodes of $\widetilde{\mathbb{G}}$ so that the endpoints of each arc in $\widetilde{\mathbb{G}}$ are assigned different colors. Each color will correspond to a distinct box, so using the minimal number of colors means using the fewest boxes. We start with the node with the highest weight. This is node a, whose weight is 3.01. Suppose we assign to a the color yellow, as indicated by the node name in a small square, as shown in Fig. 3.7. Then nodes c, e, and f cannot be colored yellow, so we color them blue, as indicated by the node name in a small circle. The remaining node in $\widetilde{\mathbb{G}}$ to be colored is node b, and it can be colored yellow.

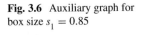

Fig. 3.6 Auxiliary graph for box size $s_1 = 0.85$

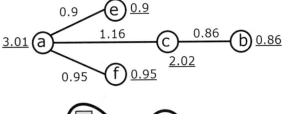

Fig. 3.7 Two boxes are required for box size s_1

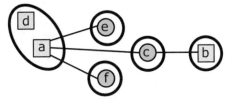

The final step in this iteration is to color to each node not in $\widetilde{\mathbb{G}}$. If node i does not appear in $\widetilde{\mathbb{G}}$ then, by construction of $\widetilde{\mathbb{G}}$, we have $d_{ij} < s_1$ for $j \in \mathbb{N}$. Thus i can be assigned to any nonempty box, so i can be assigned any previously used color. In our example, only d is not in $\widetilde{\mathbb{G}}$, and we arbitrarily color it yellow. Thus for the initial box size s_1 we require only two boxes, a yellow box containing a, b, and d, and a blue box containing c, e, and f. This concludes the iteration for the initial box size s_1.

For the second iteration, we first need the new box size s_2. The method of [69] simply takes the next smallest of the sums. Since $s_1 = \sigma(4)$, then $s_2 = \sigma(3) = 0.45$. With this new box size, we continue as before, creating a new auxiliary graph $\widetilde{\mathbb{G}}$ containing each arc (i, j) such that $d_{ij} \geq s_2$, determining the weight of each node in $\widetilde{\mathbb{G}}$, coloring the nodes of $\widetilde{\mathbb{G}}$, and then coloring the nodes not in $\widetilde{\mathbb{G}}$. Then we select $s_3 = \sigma(2)$ and continue in this manner.

This above material is one of the few sections in this survey concerning weighted networks. We now return to the study of unweighted networks, and examine other methods for box counting.

3.3 Random Sequential Node Burning

In this section we study the *Random Sequential Node Burning* method of [29] for covering $\mathbb{G}$ using radius-based boxes. For a given radius r, the procedure is as follows. Initially all nodes are uncovered (or "unburned", in the terminology of [29], i.e., not yet assigned to a box), and the box count $B_R(r)$ is initialized to 0. In each iteration, we first pick a random node n which may be covered or uncovered, but which has not previously been selected as the center node of a box. We create the new radius-based box $\mathbb{G}(n, r)$. Next we add to $\mathbb{G}(n, r)$ each uncovered node whose distance from n does not exceed r. If $\mathbb{G}(n, r)$ contains no uncovered nodes, then

Fig. 3.8 Covering using
*Random Sequential Node
Burning*

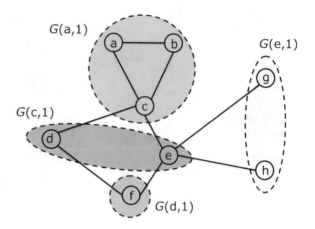

$\mathbb{G}(n, r)$ is discarded; otherwise, $B_R(r)$ is incremented by 1, and each uncovered node added to $\mathbb{G}(n, r)$ is marked as covered. This concludes an iteration of the method.

If there are still uncovered nodes, the next iteration begins by picking another random node n' which may be covered or uncovered, but which has not previously been selected as the center node of a box. We continue in this manner until all nodes are covered.

Figure 3.8, adapted from [29], illustrates the method for $r = 1$. Initially, all eight nodes are uncovered. Suppose node a is randomly selected as the center node, and $\mathbb{G}(a, 1)$ is created. Since a, b, and c are uncovered and are within one hop of a, they are placed in $\mathbb{G}(a, 1)$; nodes a, b, and c are now covered. There are still uncovered nodes. Suppose b is randomly selected as the next center node, and $\mathbb{G}(b, 1)$ is created. There are no uncovered nodes within one hop of b, and b is already covered, so we discard $\mathbb{G}(b, 1)$. Suppose c is randomly selected as the next center node, and $\mathbb{G}(c, 1)$ is created. The uncovered nodes d and e are one hop away from c, so they are placed in $\mathbb{G}(c, 1)$; nodes d, and e are now covered. There are still uncovered nodes, so suppose d is randomly selected as the next center node, and $\mathbb{G}(d, 1)$ is created. The only uncovered node one hop from d is f, so we place f in $\mathbb{G}(d, 1)$; node f is now covered. There are still uncovered nodes, so suppose e is randomly selected as the next center node, and $\mathbb{G}(e, 1)$ is created. Node g and h are uncovered and one hop from e so we place g and h in $\mathbb{G}(e, 1)$; nodes g and h are now covered. At this point there are no uncovered nodes, and the *Random Sequential Node Burning* method halts for this radius r; four boxes were created.
□

As mentioned earlier, this method can generate disconnected boxes. For example, in Fig. 3.8 there is no path in $\mathbb{G}(c, 1)$ connecting the two nodes in that box, and there is no path in $\mathbb{G}(e, 1)$ connecting the two nodes in that box. Also, a box centered at a node may not even contain the center node, e.g., $\mathbb{G}(e, 1)$ does not contain the center node e. Thus a radius-based box $\mathbb{G}(n, r)$ generated by *Random Sequential Node Burning* may fail to satisfy Definition 2.3, which would appear to cast doubt on the validity of d_B calculated by this method. However, computational results [29, 56] show that *Random Sequential Node Burning* yields the same value of d_B as obtained using other box counting methods.

It is perhaps surprising that *Random Sequential Node Burning* does not exclude a node that has been covered from being selected as a center. If *Random Sequential Node Burning* is modified to exclude a covered node from being a box center, then a power law scaling relation is not observed for the WWW [29]. However, for another fractal network studied in [29], a power law continues to hold with this modification, although the modification led to somewhat different values of $B_R(r)$.

3.4 Set Covering Formulation and a Greedy Method

The problem of computing the minimal number $B_R(r)$ of radius-based boxes of radius at most r needed to cover $\mathbb{G}$ can be formulated as a *set covering problem*, a classic combinatorial optimization problem. For simplicity we will refer to node j rather than node n_j, so $\mathbb{N} = \{1, 2, \cdots, N\}$. For a given positive integer r, let M^r be the N by N matrix defined by

$$M^r_{ij} = \begin{cases} 1 & \text{if } dist(i, j) \leq r, \\ 0 & \text{otherwise}. \end{cases} \tag{3.1}$$

(The superscript r does *not* mean the r-th power of the matrix M.) For an undirected graph, M^r is symmetric. For example, for $r = 1$, the matrix M^1 corresponding to the network of Fig. 3.9 is the same as the node-node incidence matrix of the network, namely

$$M^1 = \begin{pmatrix} 1 & 1 & 0 & 0 & 0 & 0 & 1 \\ 1 & 1 & 1 & 0 & 0 & 0 & 0 \\ 0 & 1 & 1 & 1 & 0 & 0 & 1 \\ 0 & 0 & 1 & 1 & 1 & 0 & 0 \\ 0 & 0 & 0 & 1 & 1 & 1 & 0 \\ 0 & 0 & 0 & 0 & 1 & 1 & 1 \\ 1 & 0 & 1 & 0 & 0 & 1 & 1 \end{pmatrix}.$$

For the same network and $r = 2$ we have

$$M^2 = \begin{pmatrix} 1 & 1 & 1 & 0 & 0 & 1 & 1 \\ 1 & 1 & 1 & 1 & 0 & 0 & 1 \\ 1 & 1 & 1 & 1 & 1 & 1 & 1 \\ 0 & 1 & 1 & 1 & 1 & 1 & 1 \\ 0 & 0 & 1 & 1 & 1 & 1 & 1 \\ 1 & 0 & 1 & 1 & 1 & 1 & 1 \\ 1 & 1 & 1 & 1 & 1 & 1 & 1 \end{pmatrix}.$$

Fig. 3.9 Example network
with seven nodes and eight
arcs

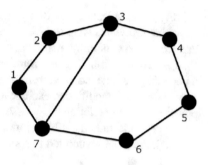

For $n = 1, 2, \cdots, N$, let x_n be the binary variable defined by

$$x_n = \begin{cases} 1 \text{ if the box centered at } n \text{ with radius } r \text{ is used in the covering of } \mathbb{G}, \\ 0 \text{ otherwise.} \end{cases}$$

The minimal number $B_R(r)$ of boxes needed to cover $\mathbb{G}$ is the optimal objective function value of the following binary integer program (an *integer program* is a linear optimization problem whose variables are restricted to be integer valued):

$$\text{minimize } \sum_{n=1}^{N} x_n \tag{3.2}$$

$$\text{subject to } \sum_{n=1}^{N} M_{jn}^r x_n \geq 1 \text{ for } j = 1, 2, \cdots, N \tag{3.3}$$

$$x_n = 0 \text{ or } 1 \text{ for } n = 1, 2, \cdots, N. \tag{3.4}$$

Here M_{jn}^r is the element in row j and column n of the matrix M^r. The objective function (3.2) is the number of center nodes used in the covering, i.e., the number of boxes used in the covering. The left hand side of constraint (3.3) is the number of boxes covering node j, so this constraint requires that each node be within distance r of at least one center node used in the covering.

To express this formulation more compactly, let $x = (x_1, x_2, \cdots, x_N)$ be the column vector of size N (that is, a matrix with N rows and one column) and let x^T be the transpose of x (so x^T is a matrix with one row and N columns). Let $\mathbf{1} = (1, 1, \cdots, 1)$ be the column vector of size N each of whose components is 1. Then the above set covering formulation can be written as

$$\text{minimize } x^T \mathbf{1}$$

$$\text{subject to } M^r x \geq \mathbf{1}$$

$$x_n = 0 \text{ or } 1 \text{ for } n = 1, 2, \cdots, N.$$

Let $\widetilde{g}(n)$ be the sum of the entries in column n of the matrix M^r, i.e., $\widetilde{g}(n) \equiv \sum_{j=1}^{N} M_{jn}^r$. Then $\widetilde{g}(n)$ is the number of nodes whose distance from node n does not exceed r. Intuitively, a node n for which $\widetilde{g}(n)$ is high (for example, a hub node) has more value in a covering of $\mathbb{G}$ than a node with for which $\widetilde{g}(n)$ is low. However, once some boxes have been selected to be in the covering, to determine which additional boxes to add to the covering, the computation of $\widetilde{g}(n)$ should consider only nodes not yet covered by any box. Therefore, given the binary vector $x \in \mathbb{R}^N$, for $n \in \mathbb{N}$ we define

$$U(x) \equiv \left\{ j \in \mathbb{N} \,\middle|\, \sum_{n=1}^{N} M_{jn}^r \, x_n = 0 \right\},$$

so if $j \in U(x)$ then node j is currently uncovered. Define

$$g(n) \equiv \sum_{1 \le j \le N, \; j \in U(x)} M_{jn}^r,$$

so $g(n)$ is the number of currently uncovered nodes that would be covered if the radius-based box $\mathbb{G}(n, r)$ centered at n were added to the covering. The $g(n)$ values are used in a greedy heuristic to solve the integer optimization problem defined by (3.2)–(3.4). Greedy methods for set covering have been known for decades, and in 1979 Chvátal [5] provided a bound on the deviation from optimality of a greedy method for set covering.

For a given r, the greedy *Maximum Excluded Mass Burning* method of [56] begins by initializing each component of the vector $x \in \mathbb{R}^N$ to zero, indicating that no node has yet been selected as the center of a radius-based box. The set Z of uncovered nodes is initialized to $\mathbb{N}$. In each iteration of *Maximum Excluded Mass Burning*, the node selected to be the center of a radius-based box is a node j for which $g(j) = \max_i \{g(i) \mid x_i = 0\}$. That is, the next center node j is a node which has not previously been selected as a center, and which, if used as the center of a box of radius r, covers the maximal number of uncovered nodes. There is no requirement that j be uncovered. In the event that more than one node yields $\max_i \{g(i) \mid x_i = 0\}$, a node can be randomly chosen, or a deterministic tie breaking rule can be utilized. We set $x_j = 1$ indicating that j has now been used as a box center. Each uncovered node i within distance r of j is removed from Z since i is now covered. If Z is now empty, we are done. Otherwise, since the newly added box will cover at least one previously uncovered node, we update $g(n)$ for each n such that $x_n = 0$ (i.e., for each node not serving as a box center). Upon termination, the estimate of $B_R(r)$ is $\sum_{n \in \mathbb{N}} x_n$. As with *Random Sequential Node Burning*, a box generated by *Maximum Excluded Mass Burning* can be disconnected [56].

Once the method has terminated, each non-center (a node n for which $x_n = 0$) is within distance r of a center (a node n for which $x_n = 1$). We may now want to assign each non-center to a center. One way to make this assignment is to arbitrarily assign each non-center n to any center c for which $dist(c, n) \le r$. Another way to make this assignment is to assign each non-center n to the closest center, breaking ties randomly. Yet another way is suggested in [56]. Although the number of nodes in each box is not needed to calculate d_B, the number of nodes in each box is required to calculate other network fractal dimensions, as will be discussed in Chaps. 8–10.

3.5   Box Burning

The *Box Burning* method [56] is a heuristic for computing the minimal number $B_D(s)$ of diameter-based boxes of size at most s needed to cover $\mathbb{G}$. Let C be the set of covered nodes, so initially $C = \emptyset$. Initialize $B_D(s) = 0$. In each iteration, a random uncovered node x is selected to be the initial occupant (i.e., the "seed") of a new box $\mathbb{G}(s)$. Since we have created a new box, we increment $B_D(s)$ by 1. We add x to C. Now we examine each node n not in C; if n is within distance $s - 1$ of each node in $\mathbb{G}(s)$ (which initially is just the seed node x), then n is added to $\mathbb{G}(s)$ and added to C. We again examine each node n not in C; if n is within distance $s - 1$ of each node in $\mathbb{G}(s)$, then n is added to $\mathbb{G}(s)$ and added to C. This continues until no additional nodes can be added to $\mathbb{G}(s)$. At this point a new box is needed, so a random uncovered node x is selected to be the seed of a new box and we continue in this manner, stopping when $C = \mathbb{N}$. We illustrate *Box Burning* for $s = 2$ using the network of Fig. 3.10.

Iteration 1 Choose f as the first seed node, create a new box containing only this node, and increment $B_D(2)$. Add node e to the box; this is allowed since the distance from e to each node in the box (which currently contains only f) does not exceed 1. No additional uncovered nodes can be added to box with nodes $\{e, f\}$; e.g., g cannot be added since its distance to e is 2.

Iteration 2 Choose b as the second seed node and create a new box containing only this node. Add node c to the box; this is allowed since the distance from c to each node in the box (which currently contains only b) is 1. No additional uncovered nodes can be added to the box with nodes $\{b, c\}$; e.g., a cannot be added since its distance to c is 2.

Fig. 3.10 Network for illustrating box burning

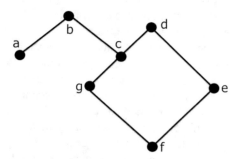

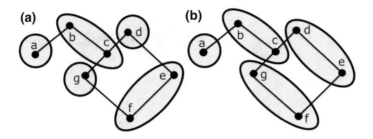

Fig. 3.11 (**a**) Box burning 2-covering and (**b**) a minimal 2-covering

Iteration 3 Choose a as the third seed node, and create a new box containing only this node. No additional uncovered nodes can be added to this box, since the only uncovered nodes are d and g, and their distance from a exceeds 1.

Iteration 4 Choose d as the fourth seed node; d can be the only occupant of this box.

Iteration 5 Finally, choose g as the fifth seed node; g can be the only occupant of this box.

We created five boxes to cover the network, which is not optimal since we can cover it using the four boxes with node sets $\{a\}$, $\{b, c\}$, $\{d, e\}$, and $\{f, g\}$. The covering created by *Box Burning* and this optimal covering are illustrated in Fig. 3.11. □

Just as the *Random Sequential Node Burning* method described above in Sect. 3.3 can generate disconnected boxes, the boxes generated by *Box Burning* can be disconnected. (So radius-based box heuristics for covering $\mathbb{G}$, as well as diameter-based box heuristics for covering $\mathbb{G}$, can generate disconnected boxes.) This is illustrated by the network of Fig. 3.12 for $s = 3$. All five nodes cannot be in the same box of diameter 2. If w is the first random seed, then the first box contains u, v, and w. If x is selected as the second random seed, the second box, containing x and y, is disconnected. The only path connecting x and y goes through a node in the first box. These two boxes form a minimal 3-covering of $\mathbb{G}$. However, for this network and $s = 3$, there is a minimal 3-covering using two connected boxes: place x, y, and u in the first box, and v and w in the second box.

While the *Box Burning* method is easy to implement, its running time is excessive. A faster *Compact Box Burning* heuristic [56] also uses diameter-based boxes. To begin, initialize $B_D(s) = 0$. Each iteration of *Compact Box Burning* processes the set U, the set of uncovered nodes, using the following steps. (*i*) Initialize $Z = \emptyset$, where Z is the set of nodes in the next box created in the covering of $\mathbb{G}$. Initialize $U = \mathbb{N}$. Increment $B_D(s)$ by 1. (*ii*) Select a random node $x \in U$; add x to Z and remove x from U, since x is now covered by the box with node set Z. (*iii*) For each node $y \in U$, if $dist(x, y) \geq s$ then remove y from U, since x and y cannot both belong to Z. (*iv*) Repeat steps (*ii*) and (*iii*) until U is empty.

Fig. 3.12 Disconnected
boxes in a minimal covering

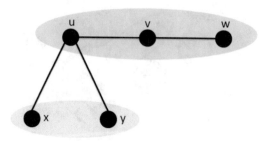

When U is empty, no more nodes can be added to Z. If $\mathbb{N}$ is empty, we are done.
If $\mathbb{N}$ is not empty, another iteration of *Compact Box Burning* is required. To start
this new iteration, first remove each node in Z from $\mathbb{N}$, since these nodes are now
covered. Now continue the iteration, starting with step (*i*).

Note that in the first iteration of *Compact Box Burning*, when we initialize $U = \mathbb{N}$
in step (*i*), the set $\mathbb{N}$ is the original set of all the nodes in $\mathbb{G}$. In subsequent iterations,
$\mathbb{N}$ is no longer the original set of all the nodes in $\mathbb{G}$, since each subsequent iteration
begins with removing each node in Z from $\mathbb{N}$. We illustrate *Compact Box Burning*
for $s = 2$, again using the network of Fig. 3.10.

Iteration 1 Initialize $B_D(2) = 0$. To start the first iteration, set $U = \mathbb{N}$, increment
$B_D(s)$ (since U is nonempty then at least one more box is needed), and create the
empty set Z. Choose f in step (*ii*). Add f to Z and remove f from U. In (*iii*),
remove nodes a, b, c, d from U, since their distance from f exceeds 1. Since U
is not empty, return to (*ii*) and choose e. Add e node to Z, remove it from U, and
remove node g from U, since its distance from e exceeds 1. Now U is empty.

Iteration 2 Since $Z = \{e, f\}$, removing these nodes from $\mathbb{N}$ yields $\mathbb{N} = \{a, b, c, d, g\}$. Since $\mathbb{N}$ is not empty, a new iteration is required. To start the second
iteration, initialize $U = \mathbb{N}$, increment $B_D(s)$, and create the empty set Z. Choose
b in step (*ii*). Add b to Z, remove b from U, and remove nodes d and g from U,
since their distance from b exceeds 1. Since U is not empty, return to step (*ii*) and
choose c. Add c to Z, remove c from U, and remove node a, since its distance from
c exceeds 1. Now U is empty.

Iteration 3 Since $Z = \{b, c\}$, removing these nodes from $\mathbb{N}$ yields $\mathbb{N} = \{a, d, g\}$.
Since $\mathbb{N}$ is not empty, a new iteration is required. To start the third iteration, initialize
$U = \mathbb{N}$, increment $B_D(s)$, and create the empty set Z. Choose a in step (*ii*). Add
a to Z, remove a from U, and remove nodes d and g since their distance from a
exceeds 1. Now U is empty.

Iteration 4 Since $Z = \{a\}$, removing a from $\mathbb{N}$ yields $\mathbb{N} = \{d, g\}$. Since $\mathbb{N}$ is not
empty, a new iteration is required. To start the fourth iteration, initialize $U = \mathbb{N}$,
increment $B_D(s)$, and create the empty set Z. Choose d in step (*ii*). Add d to Z,
remove d from U, and remove node g since its distance from d exceeds 1. Now U
is empty.

Iteration 5 Since $Z = \{d\}$, removing d from $\mathbb{N}$ yields $\mathbb{N} = \{g\}$. Since $\mathbb{N}$ is not empty, a new iteration is required. To start the fifth iteration, initialize $U = \mathbb{N}$, increment $B_D(s)$, and create the empty set Z. Choose g in step (*ii*); there is no choice here, since $\mathbb{N} = \{g\}$. Add g to Z, and remove it from U. Now U is empty.

Termination Since $\mathbb{N} = \{g\}$, removing g from $\mathbb{N}$ makes $\mathbb{N}$ empty, so we are done. For this particular network and random node selections, *Box Burning* and *Compact Box Burning* yielded the same covering using five boxes. □

Compact Box Burning, like the other heuristics we have presented, is not guaranteed to yield the minimal value $B_D(s)$. Since a random node is chosen in step (*ii*), the method is not deterministic, and executing the method multiple times, for the same network, will in general yield different values for $B_D(s)$. *Compact Box Burning* was applied in [56] to two networks, the cellular *E. coli* network and the *mbone* Internet multicast backbone network. [A *multicast* network [43] facilitates one-to-many transmissions (e.g., a basketball game is streamed to a set of geographically diverse viewers), or many-to-many transmissions (e.g., a set of geographically diverse committee members join a teleconference with both audio and video capabilities).] For both networks, the average value of $B_D(s)$ obtained by 10,000 executions of *Compact Box Burning* is up to 2% higher than the average for a greedy coloring method similar to the *Greedy Coloring* method described in Sect. 3.1. The conclusion in [56] is that *Compact Box Burning* provides results comparable with *Greedy Coloring*, but *Compact Box Burning* may be a bit simpler to implement.

Additionally, the four methods *Greedy Coloring, Random Sequential Node Burning, Maximum Excluded Mass Burning*, and *Compact Box Burning* are compared in [56]. Using 10,000 executions of each method, they find that all four methods yield the same box counting dimension d_B. However, the results for *Random Sequential Node Burning* show a much higher variance than for the other three methods, so for *Random Sequential Node Burning* it is not clear how many randomized executions are necessary, for a given r, for the average values $B_R(r)$ to stabilize.

3.6 Box Counting for Scale-Free Networks

As discussed in Sect. 3.3, *Random Sequential Node Burning* can create disconnected boxes. For example, in the example of Fig. 3.8, when c is selected as the third center node, the resulting box $\mathbb{G}(c, 1)$, which contains d and e but not c, is disconnected. Suppose we modify *Random Sequential Node Burning* so that the nodes inside each box are required to be connected within that box. With this modification, we still obtain $d_B \approx 2$ for a rectilinear lattice (i.e., a primitive square Bravais lattice) in $\mathbb{R}^2$. However, with this modification the power law scaling $B_R(r) \sim (2r + 1)^{-d_B}$ is not observed for the WWW [29].

The reason we must allow disconnected boxes to obtain a power law scaling for the WWW is that the WWW is a scale-free network. A network is *scale-free*

Fig. 3.13 Adjacent nodes
only connected through a hub

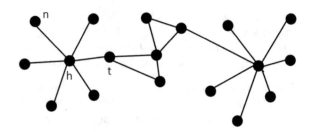

if p_k, the probability that the node degree is k, follows the power law distribution
$p_k \sim k^{-\lambda}$ for some $\lambda > 0$. For a scale-free network, a few nodes will have high
degree. Nodes with high degree (i.e., hubs) will tend to be quickly assigned to a
box, since a hub is assigned to a box whenever one of its neighbors is selected as a
center node. Once a hub is covered by a box, adjacent nodes will be disconnected
if they are connected only via the hub. This is illustrated in Fig. 3.13. Suppose we
cover this network using radius-based boxes of radius 1, and spoke n is randomly
selected as the first center node. Then hub h is added to box $\mathbb{G}(n, 1)$. If h is now
selected as a center node, all five uncovered nodes adjacent to h (e.g., t) will be
added to $\mathbb{G}(h, 1)$, even though these five nodes interconnect only through hub h.
Counting each spoke connected to h as a separate box creates too many boxes, so
Random Sequential Node Burning corrects for this phenomena by allowing nodes
in a box to be disconnected within that box. This yields a power law scaling for the
WWW.

It might be argued that allowing the nodes in a box to be disconnected within
that box violates the spirit of the Hausdorff dimension (Sect. 1.2), which covers a
geometric object $\Omega \subset \mathbb{R}^E$ by connected subsets of $\mathbb{R}^E$. However, with a change
in perspective, allowing nodes in a box to be disconnected within that box can be
viewed as using connected boxes, but allowing a node to belong to more than one
box. To see this, we revisit Fig. 3.8, but now allow overlapping boxes, as shown in
Fig. 3.14. Assume the randomly chosen centers are in the same order as before,
namely a, b, c, d, e, and again choose $r = 1$. Box $\mathbb{G}(a, 1)$ is unchanged, and
$\mathbb{G}(b, 1)$ is again discarded. However, now $\mathbb{G}(c, 1)$ also includes c, box $\mathbb{G}(d, 1)$
also contains d, and $\mathbb{G}(e, 1)$ also contains e. So c, d, and e belong to two boxes.
However, as mentioned above, the great advantage of non-overlapping boxes is that
they immediately yield a probability distribution (Chap. 8).

A very recent study [4] of almost 1000 networks, drawn from social, biological,
technological, and informational sources, concluded that, contrary to a central claim
in modern network science, scale-free networks are rare. For each of the networks,
the methodology employed was to estimate the best-fitting power law, test its
statistical plausibility, and then compare it via a likelihood ratio test to alternative
non-scale-free distributions. The study found that only 4% of the networks exhibited
the strongest possible evidence of scale-free structure, and 52% exhibited the
weakest possible evidence. The study also found that social networks are at best
weakly scale-free, while a handful of technological and biological networks can be

Fig. 3.14 Overlapping boxes

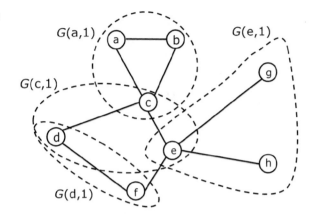

called strongly scale-free. The authors concluded that "These results undermine the universality of scale-free networks and reveal that real-world networks exhibit a rich structural diversity that will likely require new ideas and mechanisms to explain." As described in [31], this study has generated considerable discussion, with one network scientist observing that "In the real world, there is dirt and dust, and this dirt and dust will be on your data. You will never see the perfect power law."

Chapter 4
Lower Bounds on Box Counting

Consider a box counting heuristic using radius-based boxes, e.g., *Maximum Excluded Mass Burning*. There is no guarantee that the computed $B_R(r)$ is minimal or even near minimal. However, if a lower bound on $B_R(r)$ is available, we can immediately determine the deviation from optimality for the calculated $B_R(r)$. A method that provides a lower bound $B_R^L(r)$ on $B_R(r)$ is presented in [44]. The lower bound is computed by formulating box counting as an *uncapacitated facility location problem* (**UFLP**), a classic combinatorial optimization problem. This formulation provides, via the dual of the linear programming relaxation of **UFLP**, a lower bound on $B_R(r)$. The method also yields an estimate of $B_R(r)$; this estimate is an upper bound on $B_R(r)$. Under the assumption that $B_R(r) = a(2r + 1)^{-d_B}$ holds for some positive constant a and some range of r, a linear program [6], formulated using the upper and lower bounds on $B_R(r)$, provides an upper and lower bound on d_B. In the event that the linear program is infeasible, a quadratic program [18] can be used to estimate d_B.

4.1 Mathematical Formulation

Let the box radius r be fixed. For simplicity we will refer to node j rather than node n_j. Define $\mathbb{N} \equiv \{1, 2, \cdots, N\}$. Let C^r be the symmetric N by N matrix defined by

$$C_{ij}^r = \begin{cases} 0 & \text{if } dist(i, j) \leq r, \\ \infty & \text{otherwise.} \end{cases}$$

(As with the matrix M_{ij}^r defined by (3.1), the superscript r in C_{ij}^r does *not* mean the r-th power of the matrix C.) For example, for $r = 1$, the matrix C^r corresponding to the network of Fig. 4.1 is

E. Rosenberg, *A Survey of Fractal Dimensions of Networks*, SpringerBriefs in Computer Science, https://doi.org/10.1007/978-3-319-90047-6_4

Fig. 4.1 Example network
with seven nodes and
eight arcs

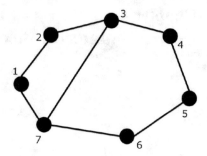

$$C^1 = \begin{pmatrix} 0 & 0 & - & - & - & - & 0 \\ 0 & 0 & 0 & - & - & - & - \\ - & 0 & 0 & 0 & - & - & 0 \\ - & - & 0 & 0 & 0 & - & - \\ - & - & - & 0 & 0 & 0 & - \\ - & - & - & - & 0 & 0 & 0 \\ 0 & - & 0 & - & - & 0 & 0 \end{pmatrix},$$

where a dash "–" is used to indicate the value ∞.
  For $j \in \mathbb{N}$, let

$$y_j = \begin{cases} 1 & \text{if the box centered at } j \text{ is used to cover } \mathbb{G}, \\ 0 & \text{otherwise.} \end{cases}$$

A given node i will, in general, be within distance r of more than one center node j
used in the covering of $\mathbb{G}$. However, we will assign each node i to exactly one node
j, and the variables x_{ij} specify this assignment. For $i, j \in \mathbb{N}$, let

$$x_{ij} = \begin{cases} 1 & \text{if } i \text{ is assigned to the box centered at } j, \\ 0 & \text{otherwise.} \end{cases}$$

With the understanding that r is fixed, for simplicity we write c_{ij} to denote element
(i, j) of the matrix C^r. The minimal network covering problem is

$$\text{minimize} \quad \sum_{j=1}^{N} y_j + \sum_{i=1}^{N} \sum_{j=1}^{N} c_{ij} x_{ij} \tag{4.1}$$

$$\text{subject to} \quad \sum_{j=1}^{N} x_{ij} = 1 \text{ for } i \in \mathbb{N} \tag{4.2}$$

$$x_{ij} \le y_j \text{ for } i, j \in \mathbb{N} \tag{4.3}$$

$$x_{ij} \geq 0 \text{ for } i, j \in \mathbb{N} \tag{4.4}$$

$$y_j = 0 \text{ or } 1 \text{ for } j \in \mathbb{N}. \tag{4.5}$$

Let **UFLP** denote the optimization problem defined by (4.1)–(4.5). Constraint (4.2) says that each node must be assigned to the box centered at some j. Constraint (4.3) says that node i can be assigned to the box centered at j only if that box is used in the covering, i.e., only if $y_j = 1$. The objective function is the sum of the number of boxes in the covering and the total cost of assigning each node to a box. Problem **UFLP** is feasible since we can always set $y_i = 1$ and $x_{ii} = 1$ for $i \in \mathbb{N}$; i.e., let each node be the center of a box in the covering. Given a set of binary values of y_j for $j \in \mathbb{N}$, since each c_{ij} is either 0 or ∞, if there is a feasible assignment of nodes to boxes then the objective function value is the number of boxes in the covering; if there is no feasible assignment for the given y_j values then the objective function value is ∞. Note that **UFLP** requires only $x_{ij} \geq 0$; it is not necessary to require x_{ij} to be binary. This relaxation is allowed since if (x, y) solves **UFLP** then the objective function value is not increased, and feasibility is maintained, if we assign each i to exactly one k (where k depends on i) such that $y_k = 1$ and $c_{ik} = 0$.

The *primal linear programming relaxation* **PLP** of **UFLP** is obtained by replacing the restriction that each y_j is binary with the constraint $y_j \geq 0$. We associate the dual variable u_i with the constraint $\sum_{j=1}^{N} x_{ij} = 1$, and the dual variable w_{ij} with the constraint $x_{ij} \geq 0$. The *dual linear program* [18] **DLP** corresponding to **PLP** is

$$\text{maximize} \quad \sum_{i=1}^{N} u_i$$

$$\text{subject to} \quad \sum_{i=1}^{N} w_{ij} \leq 1 \text{ for } j \in \mathbb{N}$$

$$u_i - w_{ij} \leq c_{ij} \text{ for } i, j \in \mathbb{N}$$

$$w_{ij} \geq 0 \text{ for } i, j \in \mathbb{N}.$$

Following [11], we set $w_{ij} = \max\{0, u_i - c_{ij}\}$ and express **DLP** using only the u_i variables:

$$\text{maximize} \quad \sum_{i=1}^{N} u_i \tag{4.6}$$

$$\text{subject to} \quad \sum_{i=1}^{N} \max\{0, u_i - c_{ij}\} \leq 1 \text{ for } j \in \mathbb{N}. \tag{4.7}$$

Let $v(UFLP)$ be the optimal objective function value of **UFLP**. Then $B_R(r) = v(UFLP)$. Let $v(PLP)$ be the optimal objective function value of the linear programming relaxation **PLP**. Then $v(UFLP) \geq v(PLP)$. Let $v(DLP)$ be the optimal objective function value of the dual linear program **DLP**. By linear programming duality theory, $v(PLP) = v(DLP)$. Define $u \equiv (u_1, u_2, \cdots, u_N)$. If u is feasible for **DLP** as defined by (4.6) and (4.7), then the dual objective function $\sum_{i=1}^{N} u_i$ satisfies $\sum_{i=1}^{N} u_i \leq v(DLP)$. Combining these relations, we have

$$B_R(r) = v(UFLP) \geq v(PLP) = v(DLP) \geq \sum_{i=1}^{N} u_i .$$

Thus $\sum_{i=1}^{N} u_i$ is a lower bound on $B_R(r)$. As described in [44], to maximize this lower bound subject to (4.7), we use the *Dual Ascent* and *Dual Adjustment* methods of [11]; see also [42].

4.2   Dual Ascent and Dual Adjustment

Call the N variables $u_1, u_2, \cdots, u_N$ the *dual variables*. The *Dual Ascent* method initializes $u = 0$ and increases the dual variables, one at a time, until constraints (4.7) prevent any further increase in any dual variable. For $i \in \mathbb{N}$, let $\mathbb{N}_i = \{j \in \mathbb{N} \mid c_{ij} = 0\}$. By definition of c_{ij}, we have $\mathbb{N}_i = \{j \mid dist(i, j) \leq r\}$. Note that $i \in \mathbb{N}_i$. From (4.7), we can increase some dual variable u_i from 0 to 1 only if $\sum_{i=1}^{N} \max\{0, u_i - c_{ij}\} = 0$ for $j \in \mathbb{N}_i$. Once we have increased u_i then we cannot increase u_k for any k such that $c_{kj} = 0$ for some $j \in \mathbb{N}_i$. This is illustrated, for $r = 1$, in Fig. 4.2, where $c_{ij_1} = c_{ij_2} = c_{ij_3} = 0$ and $c_{j_1 k_1} = c_{j_2 k_2} = c_{j_2 k_3} = 0$. Once we set $u_i = 1$, we cannot increase the dual variable associated with k_1 or k_2 or k_3.

Recalling that δ_j is the node degree of node j, if $c_{ij} = 0$ then the number of dual variables prevented by node j from increasing when we increase u_i is at least $\delta_j - 1$, where we subtract 1 since u_i is being increased from 0. In general, increasing u_i prevents approximately at least $\sum_{j \in \mathbb{N}_i} (\delta_j - 1)$ dual variables from being increased. This is approximate, since there may be arcs connecting the nodes in $\mathbb{N}_i$, e.g., there may be an arc between j_1 and j_2 in Fig. 4.2. However, we can ignore such

Fig. 4.2 Increasing u_i to 1 block other dual variable increases

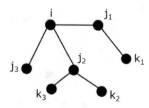

considerations since we use $\sum_{j\in\mathbb{N}_i}(\delta_j-1)$ only as a heuristic metric: we pre-process the data by ordering the dual variables in order of increasing $\sum_{j\in\mathbb{N}_i}(\delta_j-1)$. We have $\sum_{j\in\mathbb{N}_i}(\delta_j-1)=0$ only if $\delta_j=1$ for $j\in\mathbb{N}_i$, i.e., only if each node in $\mathbb{N}_i$ is a leaf node. This can occur only for the trivial case that $\mathbb{N}_i$ consists of two nodes (one of which is i itself) connected by an arc. For any other topology we have $\sum_{j\in\mathbb{N}_i}(\delta_j-1)\geq 1$. For $j\in\mathbb{N}$, define $s(j)$ to be the slack in constraint (4.7) for node j, so $s(j)=1$ if $\sum_{i=1}^{N}\max\{0,u_i-c_{ij}\}=0$ and $s(j)=0$ otherwise.

Having pre-processed the data, we run the following *Dual Ascent* procedure. This procedure is initialized by setting $u=0$ and $s(j)=1$ for $j\in\mathbb{N}$. We then examine each u_i in the sorted order and compute $\gamma\equiv\min\{s(j)\,|\,j\in\mathbb{N}_i\}$. If $\gamma=0$ then u_i cannot be increased. If $\gamma=1$ then we increase u_i from 0 to 1 and set $s(j)=0$ for $j\in\mathbb{N}_i$, since there is no longer slack in those constraints.

Figure 4.3 shows the result of applying *Dual Ascent*, with $r=1$, to *Zachary's Karate Club* network [37] , which has 34 nodes and 77 arcs. In this figure, node 1 is labelled as "v1", etc. The node with the smallest penalty $\sum_{j\in\mathbb{N}_i}(\delta_j-1)$ is node 17, and the penalty (p in the figure) is 7. Upon setting $u_{17}=1$ we have $s(17)=s(6)=s(7)=0$; these nodes are pointed to by arrows in the figure. The node with the next smallest penalty is node 25, and the penalty is 12. Upon setting $u_{25}=1$ we have $s(25)=s(26)=s(28)=s(32)=0$. The node with the next smallest penalty is node 26, and the penalty is 13. However, u_{26} cannot be increased, since $s(25)=s(32)=0$. The node with the next smallest penalty is node 12, and the penalty is 15. Upon setting $u_{12}=1$ we have $s(12)=s(1)=0$. The node with the next smallest penalty is node 27, and the penalty is 20. Upon

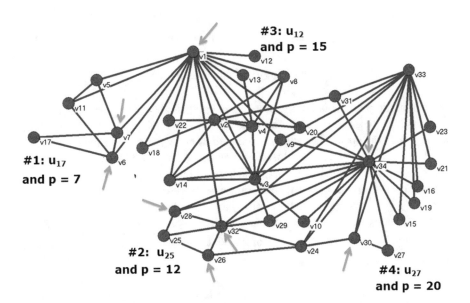

Fig. 4.3 Results of applying *Dual Ascent* to *Zachary's Karate Club* network

setting $u_{27} = 1$ we have $s(27) = s(30) = s(34) = 0$. No other dual variable can be increased, and *Dual Ascent* halts, yielding a dual objective function value of 4, which is the lower bound $B_R^L(1)$ on $B_R(1)$.

We can now calculate the upper bound $B_R^U(1)$. For $j \in \mathbb{N}$, set $y_j = 1$ if $s(j) = 0$ and $y_j = 0$ otherwise. Setting $y_j = 1$ means that the box of radius r centered at node j will be used in the covering of $\mathbb{G}$. For *Zachary's Karate Club* network, at the conclusion of *Dual Ascent* with $r = 1$ there are 12 values of j such that $s(j) = 0$; for each of these values we set $y_j = 1$.

We have shown that if u satisfies (4.7) then

$$\sum_{i=1}^{N} u_i = B_R^L(r) \leq B_R(r) \leq B_R^U(r) = \sum_{j=1}^{N} y_j.$$

If $\sum_{i=1}^{N} u_i = \sum_{j=1}^{N} y_j$ then we have found a minimal covering. If $\sum_{i=1}^{N} u_i < \sum_{j=1}^{N} y_j$ then we use a *Dual Adjustment* procedure [11] to attempt to close the gap $\sum_{j=1}^{N} y_j - \sum_{i=1}^{N} u_i$. For *Zachary's Karate Club* network, for $r = 1$ we have $\sum_{j=1}^{N} y_j - \sum_{i=1}^{N} u_i = 8$.

The *Dual Adjustment* procedure is motivated by the complementary slackness optimality conditions of linear programming. Let (x, y) be feasible for **PLP** and let (u, w) be feasible for **DLP**, where $w_{ij} = \max\{0, u_i - c_{ij}\}$. The complementary slackness conditions state that (x, y) is optimal for **PLP** and (u, w) is optimal for **DLP** if

$$y_j \left(\sum_{i=1}^{N} \max\{0, u_i - c_{ij}\} - 1 \right) = 0 \text{ for } j \in \mathbb{N} \tag{4.8}$$

$$(y_j - x_{ij}) \max\{0, u_i - c_{ij}\} = 0 \text{ for } i, j \in \mathbb{N}. \tag{4.9}$$

We can assume that x is binary, since as mentioned above, we can assign each i to a single k (where k depends on i) such that $y_k = 1$ and $c_{ik} = 0$. We say that a node $j \in \mathbb{N}$ is "open" (i.e., the box centered at node j is used in the covering of $\mathbb{G}$) if $y_j = 1$; otherwise, j is "closed." When (x, y) and u are feasible for **PLP** and **DLP**, respectively, and x is binary, constraints (4.9) have a simple interpretation: if for some i we have $u_i = 1$ then there can be at most one open node j such that $dist(i, j) \leq r$. For suppose to the contrary that $u_i = 1$ and there are two open nodes j_1 and j_2 such that $dist(i, j_1) \leq r$ and $dist(i, j_2) \leq r$. Then $c_{ij_1} = c_{ij_2} = 0$. Since x is binary, by (4.2), either $x_{ij_1} = 1$ or $x_{ij_2} = 1$. Suppose without loss of generality that $x_{ij_1} = 1$ and $x_{ij_2} = 0$. Then

$$(y_{j_1} - x_{ij_1}) \max\{0, u_i - c_{ij_1}\} = (y_{j_1} - x_{ij_1}) u_i = 0$$

but

$$(y_{j_2} - x_{ij_2}) \max\{0, u_i - c_{ij_2}\} = y_{j_2} u_i = 1,$$

so complementary slackness fails to hold. This argument is easily extended to the case where there are more than two open nodes such that $dist(i, j) \leq r$. The conditions (4.9) can also be visualized using Fig. 4.2, where $c_{ij_1} = c_{ij_2} = c_{ij_3} = 0$. If $u_i = 1$ then at most one node in the set $\{i, j_1, j_2, j_3\}$ can be open.

If $B_R^U(r) > B_R^L(r)$, we run the following *Dual Adjustment* procedure to close some nodes, and construct x, to attempt to satisfy constraints (4.9). Define

$$Y = \{j \in \mathbb{N} \mid y_j = 1\},$$

so Y is the set of open nodes. The *Dual Adjustment* procedure, which follows *Dual Ascent*, has two steps.

Step 1 For $i \in \mathbb{N}$, let $\alpha(i)$ be the "smallest" node in Y such that $c_{i,\alpha(i)} = 0$. By "smallest" node we mean the node with the smallest node index, or the alphabetically lowest node name; any similar tie-breaking rule can be used. If for some $j \in Y$ we have $j \neq \alpha(i)$ for $i \in \mathbb{N}$, then j can be closed, so we set $Y = Y - \{j\}$. In words, if the chosen method of assigning each node to a box in the covering results in the box centered at j never being used, then j can be closed.

Applying Step 1 to *Zachary's Karate Club* network with $r = 1$, using the tie-breaking rule of the smallest node index, we have, for example, $\alpha(25) = 25$, $\alpha(26) = 25$, $\alpha(27) = 27$, and $\alpha(30) = 27$. After computing each $\alpha(i)$, we can close nodes 7, 12, 17, and 28, as indicated by the bold **X** next to these nodes in Fig. 4.4. After this step, we have $Y = \{1, 6, 25, 26, 27, 30, 32, 34\}$. This step lowered the primal objective function from 12 (since originally $|Y| = 12$) to 8.

Step 2 Suppose we consider closing j, where $j \in Y$. We consider the impact of closing j on i, for $i \in \mathbb{N}$. If $j \neq \alpha(i)$ then closing j has no impact on i, since i is not assigned to the box centered at j. If $j = \alpha(i)$ then closing j is possible only if there is another open node $\beta(i) \in Y$ such that $\beta(i) \neq \alpha(i)$ and $c_{i,\beta(i)} = 0$ (i.e., if there is another open node, distinct from $\alpha(i)$, whose distance from i does not exceed r). Thus we have the rule: close j if for $i \in \mathbb{N}$ either

$$j \neq \alpha(i)$$

or

$$j = \alpha(i) \text{ and } \beta(i) \text{ exists.}$$

Once we close j and set $Y = Y - \{j\}$ we must recalculate $\alpha(i)$ and $\beta(i)$ (if it exists) for $i \in \mathbb{N}$.

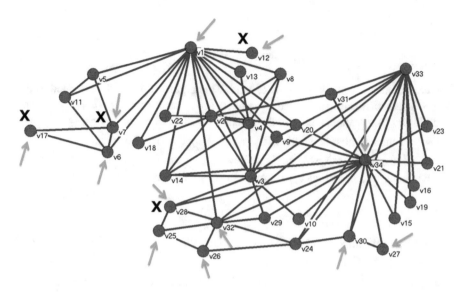

Fig. 4.4 Closing nodes in *Zachary's Karate Club* network

Applying Step 2 to *Zachary's Karate Club* network with $r = 1$, we find that, for example, we cannot close node 1, since $1 = \alpha(5)$ and $\beta(5)$ does not exist. Similarly, we cannot close node 6, since $6 = \alpha(17)$ and $\beta(17)$ does not exist. We can close node 25, since $25 = \alpha(25)$ but $\beta(25) = 26$ (i.e., we can reassign node 25 from the box centered at 25 to the box centered at 26), $25 = \alpha(26)$ but $\beta(26) = 26$, $25 = \alpha(28)$ but $\beta(28) = 34$, and $25 = \alpha(32)$ but $\beta(32) = 26$. After recomputing $\alpha(i)$ and $\beta(i)$ for $i \in \mathbb{N}$, we determine that node 26 can be closed. Continuing in this manner, we determine that nodes 27 and 30 can be closed, yielding $Y = \{1, 6, 32, 34\}$. Since now the primal objective function value and the dual objective function value are both 4, we have computed a minimal covering. When we execute *Dual Ascent* and *Dual Adjustment* for *Zachary's Karate Club* network with $r = 2$ we obtain primal and dual objective function values of 2, so again a minimal covering has been found.

4.3 Bounding the Fractal Dimension

Assume that for some positive constant a we have

$$B_R(r) = a(2r + 1)^{-d_B} . \qquad (4.10)$$

Suppose we have computed $B_R^L(r)$ and $B_R^U(r)$ for $r = 1, 2, \cdots, K$. From

$$B_R^L(r) \leq B_R(r) \leq B_R^U(r)$$

we obtain, for $r = 1, 2, \cdots, K$,

$$\log B_R^L(r) \leq \log a - d_B \log(2r + 1) \leq \log B_R^U(r). \tag{4.11}$$

The system (4.11) of $2K$ inequalities may be infeasible, i.e., it may have no solution a and d_B. If the system (4.11) is feasible, we can formulate a linear program to determine the maximal and minimal values of d_B [44]. For simplicity of notation, let the K values $\log(2r+1)$ for $r = 1, 2, \cdots, K$ be denoted by x_k for $k = 1, 2, \cdots, K$, so $x_1 = \log(3), x_2 = \log(5), x_3 = \log(7)$, etc. For $k = 1, 2, \cdots, K$, let the K values of $\log B_R^L(r)$ and $\log B_R^U(r)$ be denoted by y_k^L and y_k^U, respectively. Let $b = \log a$. The inequalities (4.11) can now be expressed as

$$y_k^L \leq b - d_B x_k \leq y_k^U.$$

The minimal value of d_B is the optimal objective function value of **BCLP** (Box Counting Linear Program):

$$\text{minimize} \quad d_B$$
$$\text{subject to} \quad b - d_B x_k \geq y_k^L \text{ for } 1 \leq k \leq K$$
$$b - d_B x_k \leq y_k^U \text{ for } 1 \leq k \leq K.$$

This linear program has only two variables, b and d_B. Let $d_B^{\min}$ and $b^{\min}$ be the optimal values of d_B and b, respectively. Now we change the objective function of **BCLP** from *minimize* to *maximize*, and let $d_B^{\max}$ and $b^{\max}$ be the optimal values of d_B and b, respectively, for the *maximize* linear program. The box counting dimension d_B, assumed to exist by (4.10), satisfies

$$d_B^{\min} \leq d_B \leq d_B^{\max}.$$

For example [44], for the much-studied *jazz* network [19], the linear program **BCLP** is feasible, and solving the *minimize* and *maximize* linear programs yields $2.11 \leq d_B \leq 2.59$.

Feasibility of **BCLP** does not imply that the box counting relationship (4.10) holds, since the upper and lower bounds might be so far apart that alternative relationships could be posited. If the linear program is infeasible, we can assert that the network does *not* satisfy the box counting relationship (4.10). Yet even if **BCLP** is infeasible, it might be so "close" to feasible that we nonetheless want to calculate d_B. When **BCLP** is infeasible, we can compute d_B using the solution of **BCQP** (box counting quadratic program), which minimizes the sum of the squared distances to the $2K$ bounds [44]:

$$\text{minimize} \quad \sum_{k=1}^{K} (u_k^2 + v_k^2)$$
$$\text{subject to} \quad u_k = (b - d_B x_k) - y_k^L \text{ for } 1 \leq k \leq K$$
$$v_k = y_k^U - (b - d_B x_k) \text{ for } 1 \leq k \leq K.$$

Chapter 5
Correlation Dimension

The correlation dimension of a geometric object in $\mathbb{R}^E$ was introduced in [22, 23].
Extending the definition to a complex network, we say that $\mathbb{G}$ has correlation
dimension d_C if the fraction $C(s)$ of nodes at a distance less than s from a random
node follows the scaling law [32, 33, 46, 57, 66]

$$C(s) \sim s^{d_C}. \tag{5.1}$$

More formally, for a positive integer s and $n \in \mathbb{N}$, define

$$C(n, s) \equiv \sum_{\substack{x \in \mathbb{N} \\ x \neq n}} I\big(s - dist(x, n)\big), \tag{5.2}$$

where $I(z) \equiv 1$ if $z > 0$ and $I(z) \equiv 0$ otherwise. Thus $C(n, s)$ is the number
of nodes, distinct from n, whose distance from n is less than s. It follows that
$C(n, s)/(N - 1)$ is the fraction of nodes, distinct from n, whose distance from n
is less than s. We have $C(n, 1) = 0$, since no node distinct from n has distance less
than 1 from n. Also, for each n and $s > \Delta$ we have $C(n, s) = N - 1$. The correlation
sum $C(s)$ is defined by

$$C(s) \equiv \frac{1}{N} \sum_{n \in \mathbb{N}} \frac{C(n, s)}{N - 1}. \tag{5.3}$$

Thus $C(s)$ is the average, over all N nodes, of the fraction $C(n, s)/(N - 1)$. It
follows that $0 \leq C(s) \leq 1$. Since $C(n, 1) = 0$ for all n then $C(1) = 0$. Note that
$C(n, 2)$ is the number of nodes at distance 1 from node n; this is δ_n, the node degree
of n. Let δ be the average node degree. From (5.3),

E. Rosenberg, *A Survey of Fractal Dimensions of Networks*, SpringerBriefs
in Computer Science, https://doi.org/10.1007/978-3-319-90047-6_5

$$C(2) = \frac{1}{N} \sum_{n \in \mathbb{N}} \frac{\delta_n}{N-1} = \frac{1}{N-1} \sum_{n \in \mathbb{N}} \frac{\delta_n}{N} = \frac{\delta}{N-1}. \tag{5.4}$$

While defining the correlation sum using (5.3) elucidates its meaning, the correlation sum is usually written as

$$C(s) = \frac{1}{N(N-1)} \sum_{n \in \mathbb{N}} \sum_{\substack{x \in \mathbb{N} \\ x \neq n}} I(s - dist(x, n)).$$

The goal is to determine if (5.1) holds for some value of d_C. The typical approach [32, 33, 57, 66] to computing d_C is to determine a range of s over which $\log C(s)$ is nearly a linear function of $\log s$, and then fit a straight line to the $(\log s, \log C(s))$ values over this range of s. The slope of this line is the estimate of the correlation dimension d_C.

In the beginning of Sect. 3.1, we observed that the problem of determining the minimal number of diameter-based boxes of size at most s needed to cover $\mathbb{G}$ is an example of the NP-hard graph coloring problem. Thus we cannot expect to compute a minimal s-covering of $\mathbb{G}$ in polynomial time. In contrast, to compute $C(s)$ for a range of s values it suffices to compute the distance between each pair of nodes. A simplistic implementation of Dijkstra's method finds the shortest path from a given source node to all other nodes in $O(N^2)$ time. Running Dijkstra N times, once for each source node, yields a time complexity of $O(N^3)$ for computing the distance between each pair of nodes. We can actually do better than $O(N^3)$ since the $N(N-1)/2$ distances can be computed in $O(N^{2.376} \log N)$ time [66]. Thus, besides the intrinsic interest in studying (5.1), computing d_C may be easier than computing the box counting dimension d_B.

The special case where $\mathbb{G}$ is a rectilinear lattice was studied in [46]. Consider a finite rectilinear lattice in $\mathbb{Z}^E$, where E is a positive integer and where $\mathbb{Z}$ denotes the integers. Assume the lattice is uniform, so each edge of the lattice contains K nodes. Thus for $E = 1$ the lattice is a chain of K nodes; for $E = 2$ it is a $K \times K$ grid, and for $E = 3$ it is a $K \times K \times K$ cube. Any definition of the correlation dimension $d_C(K)$ for such a network should satisfy two requirements. The first, the *infinite grid requirement*, is

$$\lim_{K \to \infty} d_C(K) = E. \tag{5.5}$$

Since for geometric objects $X, Y \subset \mathbb{R}^E$ any reasonable definition of dimension should satisfy the *monotonicity requirement* [12] that

$$X \subset Y \text{ implies } dimension(X) \leq dimension(Y),$$

we make the following second assertion. Let $\mathbb{G}_1$ and $\mathbb{G}_2$ be rectilinear grids in $\mathbb{Z}^E$, where each edge of $\mathbb{G}_1$ contains K_1 nodes, and each edge of $\mathbb{G}_2$ contains K_2 nodes.

Any definition of the correlation dimension should satisfy the *grid monotonicity* requirement

$$\text{if } K_1 < K_2 \text{ then } d_C(K_1) \le d_C(K_2).\tag{5.6}$$

We now present the "overall slope" formula [46] for $d_C(K)$ which satisfies both (5.5) and (5.6).

Consider first a one-dimensional grid ($E = 1$) of K nodes. As shown in [46],

$$C(s) = \frac{(2K - s)(s - 1)}{K(K - 1)}.\tag{5.7}$$

While we might hope that $d_C = 1$, the plot of $\log C(s)$ vs. $\log s$ shows that $C(s)$ does not exhibit the scaling $C(s) \sim s$. In fact, treating s as a continuous variable, $C(s)$ is strictly concave, since the second derivative is $-2/[K(K - 1)]$. Thus the least squares regression line for the ordered pairs $\big(\log s, \log C(s)\big)$ for $s = 2, 3, \cdots, K$ cannot be expected to have slope 1. However, for the one-dimensional grid there is a way to obtain exactly the desired slope of 1. Consider the "overall slope" $d_C(K)$, defined by

$$d_C(K) \equiv \frac{\log C(\Delta + 1) - \log C(2)}{\log(\Delta + 1) - \log 2} = \frac{\log C(2)}{\log[2/(\Delta + 1)]},\tag{5.8}$$

where the equality holds as $C(\Delta + 1) = 1$. This ratio represents the overall slope of the $\log C(s)$ vs. $\log s$ curve over the range $s \in [2, \Delta + 1]$. For a one-dimensional chain of K nodes we have $\Delta = K - 1$. By (5.7) we have $C(2) = 2/K$, so

$$d_C(K) = \frac{\log C(2)}{\log[2/(\Delta + 1)]} = \frac{\log(2/K)}{\log(2/K)} = 1.\tag{5.9}$$

Thus for each K the overall slope for the one-dimensional chain of size K is 1.

Now consider $d_C(K)$ for a square rectilinear grid $\mathbb{G}$ embedded in $\mathbb{Z}^2$. For $x = (x_1, x_2) \in \mathbb{Z}^2$ and $y = (y_1, y_2) \in \mathbb{Z}^2$, we have $dist(x, y) = |x_1 - y_1| + |x_2 - y_2|$. If node n is close to the boundary of $\mathbb{G}$, then the box containing all nodes a given distance from n will be truncated, as illustrated by Fig. 5.1. In Fig. 5.1a, there are four nodes in $\mathbb{G}$ whose distance to the circled node is 1. In Fig. 5.1b, there are three nodes in $\mathbb{G}$ whose distance to the circled node is 1. For the two-dimensional $K \times K$ grid, a simple expression for the overall slope can be derived [46]:

$$d_C(K) = \frac{\log[(K^2 + K)/4]}{\log[K - (1/2)]}.\tag{5.10}$$

The infinite grid requirement holds: for a $K \times K$ rectilinear grid we have

$$\lim_{K \to \infty} d_C(K) = 2.$$

Fig. 5.1 Non-truncated and truncated boxes in a 2-dimensional grid

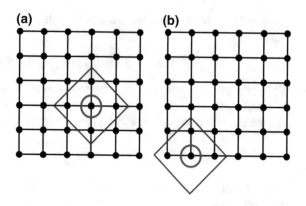

Moreover, for $K \geq 2$ we have $d_C(K) < 2$ and the monotonicity property holds: the sequence $d_C(K)$ is monotone increasing in K for $K \geq 2$. The convergence of $d_C(K)$ to the limiting value 2 is slow. Suppose we seek the value of K for which $d_C(K) = 2 - \varepsilon$, where $0 < \varepsilon < 2$. From (5.10) we have

$$\log\left(\frac{K^2 + K}{4}\right) = (2 - \varepsilon)\log[K - (1/2)]$$

which we rewrite as

$$K^2 + K = 4[K - (1/2)]^{(2-\varepsilon)}.$$

We approximate $K^2 + K$ by K^2, and $K - (1/2)$ by K, obtaining $K^2 = 4K^{(2-\varepsilon)}$, which yields $K = 4^{1/\varepsilon}$. This estimate is excellent, and shows the very slow convergence. For $\varepsilon = 1/5$ we have $K = 4^5 = 1024$ and setting $K = 1024$ in (5.10) yields $d_C(1024) \approx 1.800$; for $\varepsilon = 1/50$ we have $K = 4^{50}$ and setting $K = 4^{50}$ in (5.10) yields $d_C(4^{50}) \approx 1.980$.

For the three-dimensional $K \times K \times K$ rectilinear grid we have

$$d_C(K) = \frac{\log[(K^3 + K^2 + K)/6]}{\log[(3/2)K - 1]}. \qquad (5.11)$$

It can be proved [46] that for a $K \times K \times K$ rectilinear grid we have $d_C(K) < 3$ for $K \geq 2$, that $d_C(K)$ is monotone increasing in K for all sufficiently large K, and that $\lim_{K \to \infty} d_C(K) = 3$. The convergence of $d_C(K)$ to the limiting value 3 is also slow. Suppose we seek the value of K for which $d_C(K) = 3 - \varepsilon$, where $0 < \varepsilon < 3$. From (5.11) we have

$$\log(K^3 + K^2 + K) - \log 6 = (3 - \varepsilon)\log[(3/2)K - 1].$$

We approximate this equation by

$$\log K^3 - \log 6 = (3 - \varepsilon)\log[(3/2)K]$$

Table 5.1 Summary of overall slope results

Rectilinear grid	$d_C(K) = \dfrac{\log C(2)}{\log[2/(\Delta+1)]}$
Chain of K nodes	$d_C(K) = 1$
$K \times K$ grid	$d_C(K) = \dfrac{\log[(K^2+K)/4]}{\log[(2K-1)/2]}$
$K \times K \times K$ grid	$d_C(K) = \dfrac{\log[(K^3+K^2+K)/6]}{\log[(3K-2)/2]}$

which yields

$$K = (2/3)(81/4)^{1/\varepsilon}.$$

This approximation is also excellent. For $\varepsilon = 1$ we have $K = 13.5$ and setting $K = 13.5$ in (5.11) yields $d_C(13.5) \approx 2.060$. For $\varepsilon = 1/2$ we have $K \approx 273$ and setting $K = 273$ in (5.11) yields $d_C(273) \approx 2.502$. For $\varepsilon = 1/5$ we have $K \approx 2.270 \times 10^6$ and setting $K = 2.270 \times 10^6$ in (5.11) yields $d_C(2.270 \times 10^6) \approx 2.800$, so approximately 10^{19} nodes are required to achieve a correlation dimension of 2.8.

Rewriting (5.10) and (5.11) in a more suggestive form, the overall slope results are summarized in Table 5.1. Based on this table, it is conjectured in [46] that, for any positive integer E, the overall slope correlation dimension $d_C(K, E)$ of a uniform rectilinear grid in $\mathbb{Z}^E$, where each edge of the grid contains K nodes, is given by

$$
\begin{aligned}
d_C(K, E) &= \frac{\log[(K^E + K^{(E-1)} + \cdots + K^2 + K)/(2E)]}{\log[(EK - (E-1))/2]} \\
&= \frac{\log\left(\dfrac{K^{E+1}-K}{2E(K-1)}\right)}{\log\left(\dfrac{EK-E+1}{2}\right)}.
\end{aligned}
\tag{5.12}
$$

Expression (5.12) yields the results in Table 5.1 for $E = 1, 2, 3$.

The correlation dimension of a spatial network (a network whose nodes have natural coordinates in $\mathbb{R}^E$) was studied in [32, 33]. Their approach is to simulate random walkers who explore the network. The sequence of the coordinates of these walkers forms a time series, and the correlation dimension of the spatial network can be estimated by applying well-known techniques for computing the correlation dimension from a time series.

To conclude this chapter, we note that a generalized correlation dimension for networks was studied in [10]. For $q \leq N$, let $M_q(s)$ be the number of q-tuples of nodes such that the distance between any two nodes in a q-tuple is less than s. That is, $M_q(s)$ is the cardinality of the set

$$\{(n_{i_1}, n_{i_2}, \cdots, n_{i_q}) \text{ such that } dist(n_{i_j}, n_{i_k}) < s \text{ for } 1 \leq j < k \leq q\}.$$

This definition extends to networks the q-correlation sum defined in [20]. Since the sets are unordered, the number of sets of q nodes is given by the binomial coefficient $\binom{N}{q}$, which is upper bounded by N^q. The q-correlation function is defined as

$$C_q(s) \equiv \frac{M_q(s)}{N^q} ,$$

which for $N \gg 1$ and $q \ll N$ is approximately the fraction of all sets of q nodes with mutual distance less than s. The q-correlation dimension for $\mathbb{G}$ is defined in [10] as

$$\lim_{s \to \infty} \frac{1}{q-1} \frac{\log C_q(s)}{\log s} ,$$

so this definition assumes we can grow $\mathbb{G}$ to infinite size. For a finite complex network $\mathbb{G}$, we can estimate $C_q(s)$ using the approximation [10]

$$C_q(s) \approx \frac{1}{N} \sum_{n \in \mathbb{N}} \left(\frac{C(n, s)}{N - 1} \right)^{q-1} , \tag{5.13}$$

where $C(n, s)$ is defined by (5.2). When $q = 2$ the right hand sides of (5.13) and (5.3) are identical. In the usual fashion, for a given q the q-correlation dimension for $\mathbb{G}$ is computed by determining a range of s over which $\log C_q(s)$ is approximately linear in $\log s$; the q-correlation dimension estimate is $1/(q - 1)$ times the slope of the linear approximation.

Chapter 6
Mass Dimension for Infinite Networks

In this chapter we consider a sequence $\{\mathbb{G}_t\}_{t=1}^{\infty}$ of complex networks such that $\Delta_t \equiv diam(\mathbb{G}_t) \to \infty$ as $t \to \infty$. A convenient way to study such networks is to study how the "mass" of $\mathbb{G}_t$ scales with $diam(\mathbb{G}_t)$, where the "mass" of $\mathbb{G}_t$, which we denote by N_t, is the number of nodes in $\mathbb{G}_t$. The fractal dimension used in [73] to characterize $\{\mathbb{G}_t\}_{t=1}^{\infty}$ is

$$d_M \equiv \lim_{t \to \infty} \frac{\log N_t}{\log \Delta_t}, \tag{6.1}$$

and d_M is called the *mass dimension*. An advantage of d_M over the correlation dimension d_C is that it is sometimes much simpler to compute the network diameter than to compute $C(n, s)$ for each n and s, as is required to compute $C(s)$ using (5.3).

A procedure is presented in [73] that uses a probability p to construct a network that exhibits a transition from fractal to non-fractal behavior as p increases from 0 to 1. For $p = 0$, the network does not exhibit the small-world property and has $d_M = 2$, while for $p = 1$ the network does exhibit the small-world property and $d_M = \infty$. The construction, illustrated by Fig. 6.1, begins with $\mathbb{G}_0$, which is a single arc, and $p \in [0, 1]$. Let $\mathbb{G}_t$ be the network after t steps. The network $\mathbb{G}_{t+1}$ is derived from $\mathbb{G}_t$. For each arc in $\mathbb{G}_t$, with probability p we replace the arc with a path of three hops (introducing the two nodes c and d, as illustrated by the top branch of the figure), and with probability $1 - p$ we replace the arc with a path of four hops (introducing the three new nodes c, d, and e, as illustrated by the bottom branch of the figure). For $p = 1$, the first three generations of this construction yield the networks of Fig. 6.2. For $p = 0$, the first three generations of this construction yield the networks of Fig. 6.3. This construction builds upon the construction in [51] of (u, v) trees.

© The Author(s), under exclusive licence to Springer International Publishing AG, part of Springer Nature 2018
E. Rosenberg, *A Survey of Fractal Dimensions of Networks*, SpringerBriefs in Computer Science, https://doi.org/10.1007/978-3-319-90047-6_6

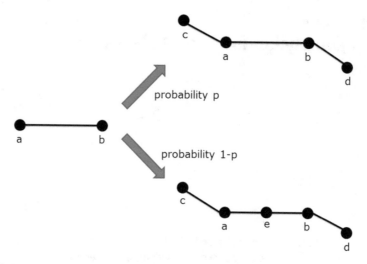

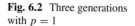

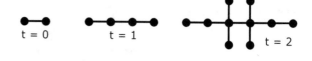

Fig. 6.1 Network that transitions from fractal to non-fractal behavior

Fig. 6.2 Three generations
with $p = 1$

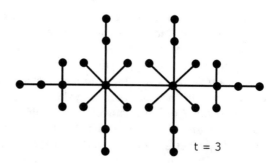

Let N_t be the expected number of nodes in $\mathbb{G}_t$, let A_t be the expected number of arcs in $\mathbb{G}_t$, and let Δ_t be the expected diameter of $\mathbb{G}_t$. The quantities N_t, A_t, and Δ_t depend on p, but for notational simplicity we omit that dependence. Since each arc is replaced by three arcs with probability p, and by four arcs with probability $1 - p$, for $t \geq 1$ we have

$$A_t = 3pA_{t-1} + 4(1 - p)A_{t-1} = (4 - p)A_{t-1}$$
$$= (4 - p)^2 A_{t-2} = \ldots = (4 - p)^t A_0 = (4 - p)^t, \qquad (6.2)$$

where the final equality follows as $A_0 = 1$, since $\mathbb{G}_0$ consists of a single arc.

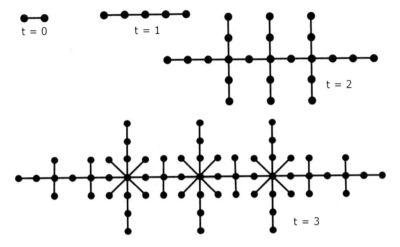

Fig. 6.3 Three generations with $p = 0$

Let x_t be the number of new nodes created in the generation of $\mathbb{G}_t$. Since each existing arc spawns two new nodes with probability p and spawns three new nodes with probability $1 - p$, from (6.2) we have

$$x_t = 2pA_{t-1} + 3(1 - p)A_{t-1} = (3 - p)A_{t-1} = (3 - p)(4 - p)^{t-1}. \quad (6.3)$$

Since $\mathbb{G}_0$ has two nodes, for $t \geq 1$ we have

$$N_t = 2 + \sum_{i=1}^{t} x_i = 2 + \sum_{i=1}^{t} (3 - p)(4 - p)^{i-1}$$

$$= 2 + (3 - p)\frac{(4 - p)^t - 1}{(3 - p)} = (4 - p)^t + 1. \quad (6.4)$$

Now we compute the diameter Δ_t of $\mathbb{G}_t$. We begin with the case $p = 1$. For this case, distances between existing node pairs are not altered when new nodes are added. At each time step, the network diameter increases by 2. Since $\Delta_0 = 1$ then $\Delta_t = 2t + 1$. Since $N_t \sim (4 - p)^t$, then the network diameter grows as the logarithm of the number of nodes, so $\mathbb{G}_t$ exhibits the small-world property for $p = 1$. From (6.1) we have $d_M = \infty$.

Now consider the case $0 \leq p < 1$. For this case, the distances between existing nodes are increased. Consider an arc in the network $\mathbb{G}_{t-1}$, and the endpoints i and j of this arc. With probability p, the distance between i and j in $\mathbb{G}_t$ is 1, and with probability $1 - p$, the distance between i and j in $\mathbb{G}_t$ is 2. The expected distance between i and j in $\mathbb{G}_t$ is therefore $p + 2(1 - p) = 2 - p$. Since each $\mathbb{G}_t$ is a tree, for $t \geq 1$ we have

$$\Delta_t = p\Delta_{t-1} + 2(1 - p)\Delta_{t-1} + 2 = (2 - p)\Delta_{t-1} + 2$$

and $\Delta_0 = 1$. This yields [73]

$$\Delta_t = \left(1 + \frac{2}{1-p}\right)(2-p)^t - \frac{2}{1-p} . \tag{6.5}$$

From (6.1), (6.4), and (6.5),

$$d_M = \lim_{t \to \infty} \frac{\log N_t}{\log \Delta_t} = \lim_{t \to \infty} \frac{\log[(4-p)^t + 1]}{\log\left[\left(1 + \frac{2}{1-p}\right)(2-p)^t - \frac{2}{1-p}\right]} = \frac{\log(4-p)}{\log(2-p)} , \tag{6.6}$$

so d_M is finite, and $\mathbb{G}_t$ does not exhibit the small-world property. For $p = 0$ we have $d_M = \log 4/\log 2 = 2$. Note that $\log(4-p)/\log(2-p) \to \infty$ as $p \to 1$.

6.1   Transfinite Fractal Dimension

A deterministic recursive construction can be used to create a self-similar network, called a (u, v)-*flower*, where u and v are positive integers [51]. By varying u and v, both fractal and non-fractal networks can be generated. The construction starts at time $t = 1$ with a cyclic graph (a ring), with $w \equiv u + v$ arcs and w nodes. At time $t + 1$, replace each arc of the time t network by two parallel paths, one with u arcs, and one with v arcs. Without loss of generality, assume $u \leq v$. Figure 6.4 illustrates three generations of a $(1, 3)$-flower. The $t = 1$ network has four arcs. To generate the $t = 2$ network, arc a is replaced by the path $\{b\}$ with one arc, and also by the path $\{c, d, e\}$ with three arcs; the other three arcs in Fig. 6.4a are similarly replaced. To generate the $t = 3$ network, arc d is replaced by the path $\{p\}$ with one arc, and also by the path $\{q, r, s\}$ with three arcs; the other fifteen arcs in Fig. 6.4b are similarly replaced. The self-similarity of the (u, v)-flowers follows from an equivalent method of construction: generate the time $t + 1$ network by making w copies of the time t network, and joining the copies at the hubs.

Let $\mathbb{G}_t$ denote the (u, v)-flower at time t. The number of arcs in $\mathbb{G}_t$ is $A_t = w^t = (u + v)^t$. The number N_t of nodes in $\mathbb{G}_t$ satisfies the recursion $N_t = wN_{t-1} - w$; with the boundary condition $N_1 = w$ we obtain [51]

$$N_t = \left(\frac{w-2}{w-1}\right)w^t + \left(\frac{w}{w-1}\right) . \tag{6.7}$$

Consider the case $u = 1$. Let Δ_t be the diameter of $\mathbb{G}_t$. It can be shown [51] that for $(1, v)$-flowers and odd v we have $\Delta_t = (v - 1)t + (3 - v)/2$ while in general, for $(1, v)$-flowers and any v,

$$\Delta_t \sim (v - 1)t . \tag{6.8}$$

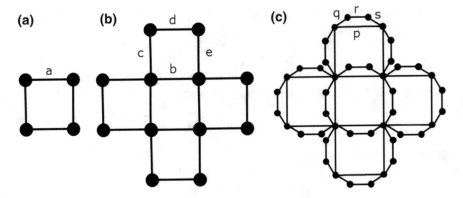

Fig. 6.4 Three generations of a $(1, 3)$-flower

Since $N_t \sim w^t$ then $\Delta_t \sim \log N_t$, so $(1, v)$-flowers enjoy the small-world property. By (6.1), (6.7), and (6.8), for $(1, v)$-flowers we have

$$d_M = \lim_{t \to \infty} \frac{\log N_t}{\log \Delta_t} = \lim_{t \to \infty} \frac{\log w^t}{\log t} = \infty, \tag{6.9}$$

so $(1, v)$-flowers have an infinite mass dimension.

We want to define a new type of fractal dimension that is finite for $(1, v)$-flowers and for other networks whose mass dimension is infinite. For $(1, v)$-flowers, from (6.7) we have

$$N_t \sim w^t = (1 + v)^t$$

as $t \to \infty$, so $\log N_t \sim t \log(1 + v)$. From (6.8) we have $\Delta_t \sim (v - 1)t$ as $t \to \infty$. Since both $\log N_t$ and Δ_t behave like a linear function of t as $t \to \infty$, but with different slopes, let d_E be the ratio of the slopes, so

$$d_E \equiv \frac{\log(1 + v)}{v - 1}. \tag{6.10}$$

From (6.10), (6.8), and (6.7), as $t \to \infty$ we have

$$d_E = \frac{t \log(1 + v)}{t(v - 1)} = \frac{\log(1 + v)^t}{t(v - 1)} = \frac{\log w^t}{t(v - 1)} \sim \frac{\log N_t}{\Delta_t}, \tag{6.11}$$

from which we obtain

$$N_t \sim e^{d_E \Delta_t}. \tag{6.12}$$

Define $\alpha_t \equiv \Delta_{t+1} - \Delta_t$. From (6.12),

$$\frac{N_{t+1}}{N_t} \sim \frac{e^{d_E \Delta_{t+1}}}{e^{d_E \Delta_t}} = e^{d_E \alpha_t}. \tag{6.13}$$

Writing $N_t = N(\Delta_t)$ for some function $N(\cdot)$, we have

$$N_{t+1} = N(\Delta_{t+1}) = N(\Delta_t + \alpha_t).$$

From this and (6.13) we have

$$N(\Delta_t + \alpha_t) \sim N(\Delta_t) e^{d_E \alpha_t}, \tag{6.14}$$

which says that, for $t \gg 1$, when the diameter increases by α_t, the number of nodes increases by a factor which is exponential in $d_E \alpha_t$. As observed in [51], in (6.14) there is some arbitrariness in the selection of e as the base of the exponential term $e^{d_E \alpha_t}$, since from (6.10) the numerical value of d_E depends on the logarithm base. If (6.14) holds as $t \to \infty$ for a sequence of self similar graphs $\{\mathbb{G}_t\}$ then d_E is called the *transfinite fractal dimension*, since this dimension "usefully distinguishes between different graphs of infinite dimensionality" [51]. Self-similar networks such as $(1, v)$-flowers whose mass dimension d_M is infinite, but whose transfinite fractal dimension d_E is finite, are called *transfinite fractal* networks, or simply *transfractals*. Thus $(1, v)$-flowers are transfractals with transfinite fractal dimension $d_E = \log(1 + v)/(v - 1)$.

Finally, consider (u, v)-flowers with $u > 1$. It can be shown [51] that $\Delta_t \sim u^t$. Using (6.7) we have

$$\lim_{t \to \infty} \frac{\log N_t}{\log \Delta_t} = \lim_{t \to \infty} \frac{\log w^t}{\log u^t} = \frac{\log(u + v)}{\log u},$$

so

$$d_M = \frac{\log(u + v)}{\log u}.$$

Since d_M is finite, these networks are fractals, not transfractals, and these networks do not enjoy the small-world property.

Chapter 7
Volume and Surface Dimensions for Infinite Networks

In this chapter we study two very early (1988) definitions [38] of the dimension of an infinite network. The two dimensions studied in this pioneering work are related to the correlation dimension d_C and mass dimension d_M. Assume that $\mathbb{G} = (\mathbb{N}, \mathbb{A})$ is an unweighted and undirected network, where both $\mathbb{N}$ and $\mathbb{A}$ are countably infinite sets. Assume the node degree of each node is finite. For $n \in \mathbb{N}$ and for each non-negative integer r, define

$$\mathbb{N}(n, r) = \{ \, x \in \mathbb{N} \, | \, dist(n, x) \leq r \, \} \, , \tag{7.1}$$

so $\mathbb{N}(n, r)$ is the set of nodes whose distance from n does not exceed r. Define

$$M(n, r) \equiv |\mathbb{N}(n, r)| \tag{7.2}$$

$$d_V^i(n) \equiv \liminf_{r \to \infty} \frac{\log M(n, r)}{\log r} \tag{7.3}$$

$$d_V^s(n) \equiv \limsup_{r \to \infty} \frac{\log M(n, r)}{\log r} \, . \tag{7.4}$$

If for some $n \in \mathbb{N}$ we have $d_V^i(n) = d_V^s(n)$, we say that $\mathbb{G}$ has *local volume dimension* $d_V(n)$ at n, where $d_V(n) \equiv d_V^i(n) = d_V^s(n)$. If also $d_V(n) = d_V$ for each n in $\mathbb{N}$, then we say that the volume dimension of $\mathbb{G}$ is d_V. The dimension d_V is called the "internal scaling dimension" in [38], but we prefer the term "volume dimension". If d_V exists then for $n \in \mathbb{N}$ we have $M(n, r) \sim r^{d_V}$ as $r \to \infty$. The volume dimension somewhat resembles the correlation dimension d_C (Chap. 5); the chief difference is that d_C is defined in terms of the correlation sum (5.3), which is an average fraction of mass, where the average is over all nodes in a finite network, while d_V is defined in terms of an infimum and supremum over all nodes. The volume dimension d_V also somewhat resembles the mass dimension d_M (Chap. 6);

© The Author(s), under exclusive licence to Springer International Publishing AG, part of Springer Nature 2018
E. Rosenberg, *A Survey of Fractal Dimensions of Networks*, SpringerBriefs in Computer Science, https://doi.org/10.1007/978-3-319-90047-6_7

the chief difference is that d_M is defined in terms of the network diameter and all the nodes in the network, while d_V is defined in terms of the number of nodes within radius r of node n.

The second definition of a network dimension proposed in [38] is based on the boundary (i.e., surface) of $N(n, r)$, rather than on the entire neighborhood $N(n, r)$. For $n \in N$ and for each non-negative integer r, define

$$\partial N(n, r) = \{ x \in N \mid dist(n, x) = r \} , \tag{7.5}$$

so $\partial N(n, r)$ is the set of nodes whose distance from n is exactly r. For each n we have

$$\partial N(n, r) = N(n, r) - N(n, r - 1)$$

and $\partial N(n, 0) = \{n\}$. Define

$$d_U^i(n) \equiv \liminf_{r \to \infty} \left(\frac{\log |\partial N(n, r)|}{\log r} + 1 \right) \tag{7.6}$$

$$d_U^s(n) \equiv \limsup_{r \to \infty} \left(\frac{\log |\partial N(n, r)|}{\log r} + 1 \right) . \tag{7.7}$$

If $d_U^i(n) = d_U^s(n)$, we say that G has *surface dimension* $d_U(n)$ at n, where $d_U(n) \equiv d_U^i(n) = d_U^s(n)$. If for each n in N we have $d_U(n) = d_U$, then we say that the surface dimension of G is d_U. This dimension is called the "connectivity dimension" in [38], since "it reflects to some extent the way the node states are interacting with each other over larger distances via the various bond sequences connecting them", but we prefer the term "surface dimension". If d_U exists, then for $n \in N$ we have $|\partial N(n, r)| \sim r^{d_U - 1}$ as $r \to \infty$. The volume dimension d_V is "rather a mathematical concept and is related to well known dimensional concepts in fractal geometry", while the surface dimension d_U "seems to be a more physical concept as it measures more precisely how the graph is connected and how nodes can influence each other". The values of d_V and d_U are identical for "generic" networks, but are different on certain "exceptional" networks [38].

Example 7.1 Let G be the infinite graph whose nodes lie on a straight line at locations $\cdots, -3, -2, -1, 0, 1, 2, 3, \cdots$. Then $M(n, r) = 2r + 1$ for each n and r, so $d_V = \lim_{r \to \infty} \log(2r + 1)/\log r = 1$. As for the surface dimension, we have $|\partial N(n, r)| = 2$ for each n and r, so $d_U = \lim_{r \to \infty}[(\log 2/\log r) + 1] = 1$. □

Example 7.2 Let G be the infinite two-dimensional rectilinear lattice whose nodes have integer coordinates, and where node (i, j) is adjacent to $(i - 1, j)$, $(i + 1, j)$, $(i, j - 1)$, and $(i, j + 1)$. Using the L_1 (i.e., Manhattan) metric, the distance from the origin to node $n = (n_1, n_2)$ is $|n_1| + |n_2|$. For integer $r \geq 1$, the number $|\partial N(n, r)|$

of nodes at a distance r from a given node n is $4r$ (see [54]), so

$$|\mathbb{N}(n, r)| = 1 + 4 \sum_{i=1}^{r} i = 1 + 4r(r + 1)/2 = 2r^2 + 2r + 1.$$

Hence

$$d_U = \lim_{r \to \infty} [(\log 4r / \log r) + 1] = 2$$

$$d_V = \lim_{r \to \infty} \log(2r^2 + 2r + 1) / \log r = 2. \quad \square$$

A construction and corresponding analysis in [38] generates an infinite network for which d_V exists but d_U does not exist. Thus the existence of the volume dimension d_V, and its numerical value, do not provide much information about the behavior of $|\partial\mathbb{N}(n, r)|$, although the inequality $\limsup_{r \to \infty} \log |\partial\mathbb{N}(n, r)| / \log r \leq d_V(n)$ is valid for all n. On the other hand, the existence of the surface dimension at n does imply the existence of the volume dimension at n, and the equality of these values: it is proved in [38] that if $d_U(n)$ exists and if $d_U(n) > 1$ then $d_V(n)$ exists and $d_V(n) = d_U(n)$. It is also proved that if $n \in \mathbb{N}$ and h is a positive number, then the insertion of arcs between arbitrarily many pairs of nodes (x, y), subject to the constraint that $dist(x, y) \leq h$, does not change $d_V^i(n)$ and $d_V^s(n)$.

Let d be an arbitrary number such that $1 < d \leq 2$. It is shown in [38] how to construct a conical graph such that $d_V(n^\star) = d$, where $n^\star$ is a specially chosen node. Graphs with higher volume dimension can be constructed using a nearly identical method. The construction is illustrated in Fig. 7.1 for the choice $d = 5/3$. There are two types of nodes in this figure, black circles and grey squares. Both types of nodes are nodes in $\mathbb{G}$, but we distinguish these two types to facilitate calculating $d_V(n^\star)$. From the figure we see that boxes connect level m to level $m + 1$, where the lower left and lower right corner of each box is a grey square, and the upper left and upper right corner of each box is a black circle. (These boxes have nothing to do with the boxes used to compute the box counting dimension.) For $m \geq 2$, we construct $\lfloor (2m - 1)^{d-1} \rfloor$ boxes connecting level m to level $m + 1$, so the number of grey squares at level m is $1 + \lfloor (2m - 1)^{d-1} \rfloor$. Each grey square at level m has distance $2m - 1$ to node $n^\star$. Thus, for $m \geq 2$, the set of grey squares at level m is the set $\partial\mathbb{N}(n^\star, 2m - 1)$, and

$$|\partial\mathbb{N}(n^\star, 2m - 1)| = 1 + \lfloor (2m - 1)^{d-1} \rfloor.$$

To compute $d_V(n^\star)$, we first determine $d_U(n^\star)$, the surface dimension at $n^\star$. Since

$$(2m - 1)^{d-1} \leq 1 + \lfloor (2m - 1)^{d-1} \rfloor \leq 1 + (2m - 1)^{d-1}$$

Fig. 7.1 Construction of a
conical graph for which
$d_V(n^*) = 5/3$

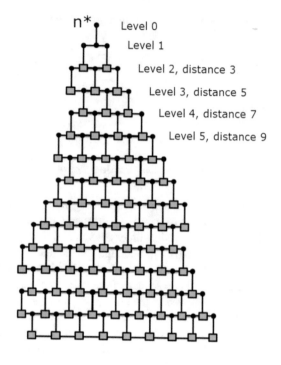

then

$$\lim_{m \to \infty} \frac{\log (2m-1)^{d-1}}{\log (2m-1)} \leq \lim_{m \to \infty} \frac{\log |\partial \mathbb{N}(n^*, 2m-1)|}{\log (2m-1)} \leq \lim_{m \to \infty} \frac{\log \left(1 + (2m-1)^{d-1}\right)}{\log (2m-1)} .$$

Since the first and third limits are both equal to $d - 1$, then so is the second limit.
That is,

$$\lim_{m \to \infty} \frac{\log |\partial \mathbb{N}(n^*, 2m-1)|}{\log (2m-1)} = d - 1 . \tag{7.8}$$

At this point we have shown that the limit exists along the subsequence of distances
$\{r_m\}_{m=1}^{\infty}$, where $r_m \equiv 2m - 1$. Since convergence of a subsequence does not imply
convergence of the entire sequence, it must be shown that the limit exists for all
sufficiently large distances, not just along a subsequence of distances. The proof is
provided in [38], along with other remaining details, and for the above conical graph
construction we have $d_V = d$.

Chapter 8
Information Dimension

The information dimension d_I of a network [47, 70] extends the concept of the information dimension of a probability distribution [2, 13, 50, 52]. Recall that for box size s, an s-covering of $\mathbb{G}$ was defined by Definition 2.1, and that an s-covering is minimal if it uses the fewest possible number of boxes. Let $\mathscr{B}(s)$ be a minimal s-covering of $\mathbb{G}$. Let $N_j(s)$ be the number of nodes of $\mathbb{G}$ contained in box $B_j \in \mathscr{B}(s)$. We obtain a set of box probabilities $\{p_j(s)\}_{B_j \in \mathscr{B}(s)}$ by defining $p_j(s) \equiv N_j(s)/N$. Define the entropy $H(s)$ by

$$H(s) \equiv - \sum_{B_j \in \mathscr{B}(s)} p_j(s) \log p_j(s) . \tag{8.1}$$

The information dimension d_I for the complex network $\mathbb{G}$ is defined in [70] by $d_I \equiv - \lim_{s \to 0} H(s)/\log s$. However, this definition is not computationally useful, since the distance between each pair of nodes is at least 1. Moreover, we cannot use the value $s = 1$, since then the denominator of the fraction is zero, while for $s \geq 2$ we have $H(s) > 0$ and $\log s > 0$, which implies $-H(s)/\log s < 0$ and thus $d_I < 0$. A computationally useful definition of d_I would, in the spirit of Definition 2.4 of d_B, require $H(s)$ to be approximately linear in $\log s$, i.e.,

$$H(s) \approx -d_I \log (s/\Delta) + c \tag{8.2}$$

for some constant c and some range of s. For example, using (8.2), it can be shown [47] that for a three-dimensional cubic rectilinear lattice of K^3 nodes, where K is the number of nodes on an edge of the cube, for $K \gg 1$ we have $d_I = 3$.

Example 8.1 Figure 8.1 illustrates a "hub and spoke with a tail" topology with N nodes. The spokes are the $N - K$ nodes $K+1, K+2, \cdots, K+N$ that are connected

E. Rosenberg, *A Survey of Fractal Dimensions of Networks*, SpringerBriefs in Computer Science, https://doi.org/10.1007/978-3-319-90047-6_8

Fig. 8.1 "Hub and spoke with a tail" network

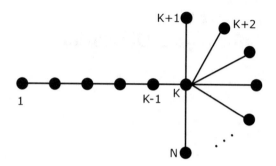

Fig. 8.2 Two minimal 3-coverings and a minimal 2-covering for the *chair* network

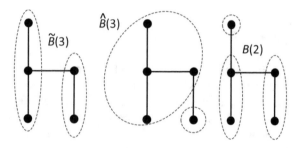

to node K. The diameter Δ of this network is K. Assuming $s \ll N$, $s \ll K$ and $s \ll N - K$, it is shown in [47] that

$$H(s) \approx -\left[\left(\frac{N-K}{N}\right)\log\left(\frac{N-K}{N}\right) + \frac{K}{N}\log\frac{K}{N}\right] - \frac{K}{N}\log\frac{s}{K}. \tag{8.3}$$

Setting

$$c = -\left[\left(\frac{N-K}{N}\right)\log\left(\frac{N-K}{N}\right) + \frac{K}{N}\log\frac{K}{N}\right],$$

from (8.3) we have

$$H(s) \approx -(K/N)\log(s/K) + c.$$

Since $\Delta = K$, from (8.2) we have $d_I = K/N$. When $N = K + 1 \gg 1$, the network is a long chain (i.e., a rectilinear lattice in one dimension), for which $d_I \approx 1$. When $N \gg K = 2$, the network is a pure hub and spoke network, for which $d_I \approx 0$. $\quad\square$

Example 8.2 Consider the five node *chair* network of Fig. 8.2. Using the minimal 2-covering $\mathscr{B}(2)$ and the minimal 3-covering $\widehat{\mathscr{B}}(3)$, from (8.2) we obtain

$$d_I = \left(H(2) - H(3)\right)/(\log 3 - \log 2) = (4/5)\log 2/\log(3/2) \approx 1.37.$$

However, using the minimal 2-covering $\mathscr{B}(2)$ and the other minimal 3-covering $\widetilde{\mathscr{B}}(3)$, from (8.2) we obtain

$$d_I = \big(H(2) - H(3)\big)/(\log 3 - \log 2) = \left(\frac{3}{5}\log 3 - \frac{2}{5}\log 2\right)\Big/\log(3/2) \approx 0.94\,.$$

□

Thus the value of d_I depends in general on which minimal covering is selected. This indeterminacy can be eliminated by using maximal entropy minimal coverings, introduced in [47].

Definition 8.1 The minimal s-covering $\mathscr{B}(s)$ of $\mathbb{G}$ is a *maximal entropy minimal covering* of $\mathbb{G}$ if the entropy $H(s)$ of $\mathscr{B}(s)$ is no less than the entropy of any other minimal s-covering of $\mathbb{G}$. □

For Example 8.2, the minimal 3-covering $\widehat{\mathscr{B}}(3)$ has entropy

$$-\left(\frac{4}{5}\log\frac{4}{5} + \frac{1}{5}\log\frac{1}{5}\right) \approx 0.217$$

while the minimal 3-covering $\widetilde{\mathscr{B}}(3)$ has entropy

$$-\left(\frac{3}{5}\log\frac{3}{5} + \frac{2}{5}\log\frac{2}{5}\right) \approx 0.292\,,$$

so $\widetilde{\mathscr{B}}(3)$ is a maximal entropy minimal 3-covering. The justification for this maximal entropy definition is the argument of Jaynes [26] that the significance of Shannon's information theory is that there is a unique, unambiguous criterion for the uncertainty associated with a discrete probability distribution, and this criterion agrees with our intuitive notion that a flattened distribution possesses more uncertainty than a sharply peaked distribution. Jaynes argues that when making physical inferences with incomplete information about the underlying probability distribution, we must use that probability distribution which maximizes the entropy subject to the known constraints. The definition of d_I provided by (8.2) can now be replaced with the following definition [47].

Definition 8.2 The network $\mathbb{G}$ has the information dimension d_I if for some constant c and for some range of s we have

$$H(s) \approx -d_I \log(s/\Delta) + c\,, \tag{8.4}$$

where $H(s)$ is given by (8.1), $p_j(s) = N_j(s)/N$ for $B_j \in \mathscr{B}(s)$, and $\mathscr{B}(s)$ is a maximal entropy minimal s-covering. □

With this definition, for the network of Fig. 8.2 we have $d_I = 0.94$. Procedure 8.1 shows how any box counting method for $\mathbb{G}$ can easily be modified to generate a maximal entropy minimal s-covering. Aside from using (8.1) to compute $H(s)$ from $\mathscr{B}(s)$, computing a maximal entropy minimal covering is no more computationally burdensome than computing a minimal covering.

Procedure 8.1 Let $\mathscr{B}_{\min}(s)$ be the best s-covering obtained over all executions of whatever box counting method is utilized. Suppose we have executed box counting some number of times, and stored $\mathscr{B}_{\min}(s)$ and $H_{\max}(s)$, where $H_{\max}(s)$ is the current estimate of the maximal entropy of a minimal s-covering. Now suppose we execute box counting again, and generate a new s-covering $\mathscr{B}(s)$ using $B(s)$ boxes. Using (8.1), compute the entropy H of the s-covering $\mathscr{B}(s)$. If $B(s) < B_{\min}(s)$, or if $B(s) = B_{\min}(s)$ and $H > H_{\max}(s)$, then set $\mathscr{B}_{\min}(s) = \mathscr{B}(s)$ and $H_{\max}(s) = H$. □

Example 8.3 Table 8.1 and Fig. 8.3 show the results after 1000 executions of *Random Sequential Node Burning* (Sect. 3.3) are applied to an Erdos-Rényi random network, with 100 nodes, 250 arcs, and diameter 6. To show the range of entropy over all minimal coverings, let $H_{\min}(s)$ be the *minimal* entropy, taken over all s-coverings for which $B(s) = B_{\min}(s)$. Over the range $1 \leq r \leq 3$ (which, using $s = 2r + 1$, translates to box sizes of 3, 5, and 7), the best linear fits of $\log B_{\min}(s)$ and $H_{\max}(s)$ vs. $\log s$ yield $d_B = 3.17$ and $d_I = 2.99$, so $d_I < d_B$. The plot of $H_{\max}(s)$ vs. $\log s$ (solid blue line, triangular markers) is substantially more linear than the plot of $H_{\min}(s)$ vs. $\log s$ (dashed red line, square markers), so the use of maximal entropy minimal coverings enhanced the ability to compute d_I. Redoing this analysis with only 100 executions of *Random Sequential Node Burning* yielded

Table 8.1 Results for Erdos-Rényi network

r	$B_{\min}(s)$	$H_{\min}(s)$	$H_{\max}(s)$
1	31	3.172	3.246
2	9	1.676	1.951
3	2	0.168	0.680
4	1	0	0

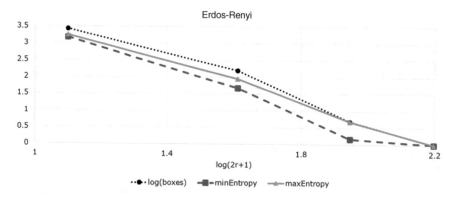

Erdos-Renyi

$\cdots\bullet\cdots$log(boxes) $\blacksquare$minEntropy $\blacktriangle$maxEntropy

Fig. 8.3 Results for Erdos-Rényi network

the quite different results $d_B = 2.71$ and $d_I = 2.72$, so the number of executions of *Random Sequential Node Burning* (or whichever box counting method is utilized) should be sufficiently high so that the results have converged.   □

Chapter 9
Generalized Dimensions

A multifractal is a fractal that cannot be characterized by a single fractal dimension such as the box counting dimension. The infinite number of fractal dimensions needed in general to characterize a multifractal are known as generalized dimensions. Generalized dimensions of geometric multifractals were proposed independently in 1983 by Grassberger [20] and by Hentschel and Procaccia [25]. They have been intensely studied (e.g., [21, 40, 61]) and widely applied (e.g., [39, 59]). Given N points from a geometric multifractal, e.g., the strange attractor of a dynamical system [9, 41], the generalized dimension D_q defined in [20, 25] is computed from a set of box sizes. For box size s, we cover the N points with a grid of boxes of linear size s, compute the fraction $p_j(s)$ of the N points in box B_j of the grid, discard any box for which $p_j(s) = 0$, and compute the partition function value

$$Z_q(\mathscr{B}(s)) \equiv \sum_{B_j \in \mathscr{B}(s)} [p_j(s)]^q , \qquad (9.1)$$

where $\mathscr{B}(s)$ is the set of non-empty grid boxes, of linear size s, used to cover the N points. For $q \geq 0$ and $q \neq 1$, the generalized dimension D_q defined in [20, 25] of the geometric multifractal is

$$D_q \equiv \frac{1}{q-1} \lim_{s \to 0} \frac{\log Z_q(\mathscr{B}(s))}{\log s} . \qquad (9.2)$$

When $q = 0$, this computation yields the box counting dimension d_B, so $D_0 = d_B$. When $q = 1$, after applying L'Hôpital's rule we obtain the information dimension d_I [13], so $D_1 = d_I$. When $q = 2$, we obtain the correlation dimension d_C [23], so $D_2 = d_C$.

Generalized dimensions of a complex network were studied in [15, 34, 48, 49, 58, 67, 68]. Several of these studies employ the *sandbox method*, which we discuss

© The Author(s), under exclusive licence to Springer International Publishing AG, part of Springer Nature 2018
E. Rosenberg, *A Survey of Fractal Dimensions of Networks*, SpringerBriefs in Computer Science, https://doi.org/10.1007/978-3-319-90047-6_9

at the end of this chapter. The method of [67] for computing D_q for $\mathbb{G}$ is the following. For a range of s, compute a minimal s-covering $\mathscr{B}(s)$. For $B_j \in \mathscr{B}(s)$, define $p_j(s) \equiv N_j(s)/N$, where $N_j(s)$ is the number of nodes in B_j. For $q \in \mathbb{R}$, use (9.1) to compute $Z_q(\mathscr{B}(s))$. (In [67], which uses a randomized box counting heuristic, $Z_q(\mathscr{B}(s))$ is the average partition function value, averaged over 200 random orderings of the nodes.) Typically, D_q is computed only for a small set of q values, e.g., integer q in $[0, 10]$ or integer q in $[-10, 10]$. Then $\mathbb{G}$ has the generalized dimension D_q (for $q \neq 1$) if for some constant c and for some range of s we have

$$\log Z_q(\mathscr{B}(s)) \approx (q - 1)D_q \log(s/\Delta) + c. \tag{9.3}$$

However, as shown in [48], this definition is ambiguous, since different minimal s-coverings can yield different values of D_q.

Example 9.1 Consider again the *chair* network of Fig. 8.2, which shows two minimal 3-coverings and a minimal 2-covering. Choosing $q = 2$, for the covering $\widetilde{\mathscr{B}}(3)$ from (9.1) we have $Z_2(\widetilde{\mathscr{B}}(3)) = (\frac{3}{5})^2 + (\frac{2}{5})^2 = \frac{13}{25}$, while for $\widehat{\mathscr{B}}(3)$ we have $Z_2(\widehat{\mathscr{B}}(3)) = (\frac{4}{5})^2 + (\frac{1}{5})^2 = \frac{17}{25}$. For $\mathscr{B}(2)$ we have $Z_2(\mathscr{B}(2)) = 2(\frac{2}{5})^2 + (\frac{1}{5})^2 = \frac{9}{25}$. If we use $\widetilde{\mathscr{B}}(3)$ then from (9.3) and the range $s \in [2, 3]$ we obtain

$$D_2 = \left(\log \frac{13}{25} - \log \frac{9}{25}\right)/(\log 3 - \log 2) \approx 0.907.$$

If instead we use $\widehat{\mathscr{B}}(3)$ and the same range of s we obtain

$$D_2 = \left(\log \frac{17}{25} - \log \frac{9}{25}\right)/(\log 3 - \log 2) \approx 1.569.$$

Thus the method of [67] can yield different values of D_2 depending on the minimal covering selected. $\square$

To devise a computationally efficient method for selecting a unique minimal covering, first consider the maximal entropy criterion described in Chap. 8. It is well known that entropy is *maximized* when all the probabilities are equal. A partition function is *minimized* when the probabilities are equal. To formalize this idea, for integer $J \geq 2$, let $\mathbf{P}(\mathbf{q})$ denote the continuous optimization problem: minimize $\sum_{j=1}^{J} p_j^q$ subject to $\sum_{j=1}^{J} p_j = 1$ and $p_j \geq 0$ for each j. It is proved in [48] that for $q > 1$, the solution of $\mathbf{P}(\mathbf{q})$ is $p_j = 1/J$ for each j. Applying this result to $\mathbb{G}$, minimizing $Z_q(\mathscr{B}(s))$ over all minimal s-coverings of $\mathbb{G}$ yields a minimal s-covering for which all the probabilities $p_j(s)$ are, to the extent possible, equalized. Since $p_j(s) = N_j(s)/N$, equal box probabilities means that all boxes in the minimal s-covering have the same number of nodes. The following definition [48] of an (s, q) minimal covering, for use in computing D_q, is analogous to the definition in [47] of a maximal entropy minimal s-covering, for use in computing d_I.

Definition 9.1 For $q \in \mathbb{R}$, the covering $\mathscr{B}(s)$ of $\mathbb{G}$ is an (s, q) minimal covering if (i) $\mathscr{B}(s)$ is a minimal s-covering and (ii) for any other minimal s-covering $\widetilde{\mathscr{B}}(s)$ we have $Z_q(\mathscr{B}(s)) \leq Z_q(\widetilde{\mathscr{B}}(s))$. □

It is easy to modify any box counting method (in a manner analogous to Procedure 8.1) to compute an (s, q) minimal covering for a given s and q. However, this approach to eliminating ambiguity in the computation of a minimal s-covering is not particularly attractive, since it requires computing an (s, q) minimal covering for each value of q for which we wish to compute D_q. A better approach to resolving this ambiguity is to compute a lexico minimal summary vector [48], which summarizes an s-covering $\mathscr{B}(s)$ by the point $x \in \mathbb{R}^J$, where $J \equiv B(s)$, where $x_j = N_j(s)$ for $1 \leq j \leq J$, and where $x_1 \geq x_2 \geq \cdots \geq x_J$. (We use *lexico* instead of the longer *lexicographically*.) The vector x does not specify all the information in $\mathscr{B}(s)$; in particular, $\mathscr{B}(s)$ specifies exactly which nodes belong to each box, while x specifies only the number of nodes in each box. The notation $x = \sum \mathscr{B}(s)$ signifies that x summarizes the s-covering $\mathscr{B}(s)$ and that $x_1 \geq x_2 \geq \cdots \geq x_J$. For example, if $N = 37$, $s = 3$, and $B(3) = 5$, we might have $x = \sum \mathscr{B}(3)$ for $x = (18, 7, 5, 5, 2)$. However, we cannot have $x = \sum \mathscr{B}(3)$ for $x = (7, 18, 5, 5, 2)$ since the components of x are not ordered correctly. If $x = \sum \mathscr{B}(s)$ then each x_j is positive, since x_j is the number of nodes in box B_j. The vector $x = \sum \mathscr{B}(s)$ a called a *summary* of $\mathscr{B}(s)$. By "x is a summary" we mean x is a summary of $\mathscr{B}(s)$ for some $\mathscr{B}(s)$. For $x(s) = \sum \mathscr{B}(s)$ and $q \in \mathbb{R}$, define

$$Z(x(s), q) \equiv \sum_{B_j \in \mathscr{B}(s)} \left(\frac{x_j(s)}{N} \right)^q. \tag{9.4}$$

Thus for $x(s) = \sum \mathscr{B}(s)$ we have $Z(x(s), q) = Z_q(\mathscr{B}(s))$, where $Z_q(\mathscr{B}(s))$ is defined by (9.1).

Let $x \in \mathbb{R}^K$ for some positive integer K. Let $right(x) \in \mathbb{R}^{K-1}$ be the point obtained by deleting the first component of x. For example, if $x = (18, 7, 5, 5, 2)$ then $right(x) = (7, 5, 5, 2)$. Similarly, we define $right^2(x) \equiv right(right(x))$, so $right^2(7, 7, 5, 2) = (5, 2)$. Let $u \in \mathbb{R}$ and $v \in \mathbb{R}$ be numbers. We say that $u \succeq v$ (in words, u is *lexico* greater than or equal to v) if ordinary inequality holds, that is, $u \succeq v$ if $u \geq v$. Thus $6 \succeq 3$ and $3 \succeq 3$. Now let $x \in \mathbb{R}^K$ and $y \in \mathbb{R}^K$. We define lexico inequality recursively: we say that $y \succeq x$ if either (i) $y_1 > x_1$ or (ii) $y_1 = x_1$ and $right(y) \succeq right(x)$. For example, for $x = (9, 6, 5, 5, 2)$, $y = (9, 6, 4, 6, 2)$, and $z = (8, 7, 5, 5, 2)$, we have $x \succeq y$ and $x \succeq z$ and $y \succeq z$.

Definition 9.2 Let $x = \sum \mathscr{B}(s)$. Then x is *lexico minimal* if (i) $\mathscr{B}(s)$ is a minimal s-covering and (ii) if $\widetilde{\mathscr{B}}(s)$ is a minimal s-covering distinct from $\mathscr{B}(s)$ and $y = \sum \widetilde{\mathscr{B}}(s)$ then $y \succeq x$. □

The following two theorems are proved in [48].

Theorem 9.1 *For each s there is a unique lexico minimal summary.*

Theorem 9.2 *Let $x = \sum \mathscr{B}(s)$. If x is lexico minimal then $\mathscr{B}(s)$ is (s, q) minimal for all sufficiently large q.*

Analogous to Procedure 8.1, Procedure 9.1 below shows how, for a given s, the lexico minimal $x(s)$ can be computed by a simple modification of whatever box counting method is used to compute a minimal s-covering.

Procedure 9.1 Let $\mathscr{B}_{min}(s)$ be the best s-covering obtained over all executions of whatever box counting method is utilized. Suppose we have executed box counting some number of times, and stored $\mathscr{B}_{min}(s)$ and $x_{min}(s) = \sum \mathscr{B}_{min}(s)$, so $x_{min}(s)$ is the current best estimate of a lexico minimal summary vector. Now suppose we execute box counting again, and generate a new s-covering $\mathscr{B}(s)$ using $B(s)$ boxes. Let $x = \sum \mathscr{B}(s)$. If $B(s) < B_{min}(s)$, or if $B(s) = B_{min}(s)$ and $x_{min}(s) \succeq x$, then set $\mathscr{B}_{min}(s) = \mathscr{B}(s)$ and $x_{min}(s) = x$. □

Procedure 9.1 shows that the only additional steps, beyond the box counting method itself, needed to compute $x(s)$ are lexicographic comparisons, and no evaluations of the partition function $Z_q(\mathscr{B}(s))$ are required. By Theorems 9.1 and 9.2, the summary vector $x(s)$ is unique and also "optimal" (i.e., (s, q) minimal) for all sufficiently large q. Thus an attractive way to resolve ambiguity in the choice of minimal s-coverings is to compute $x(s)$ for a range of s and use the $x(s)$ vectors to compute D_q, using Definition 9.3 below.

Definition 9.3 For $q \neq 1$, the complex network $\mathbb{G}$ has the generalized dimension D_q if for some constant c and for some range of s we have

$$\log Z(x(s), q) \approx (q - 1) D_q \log(s/\Delta) + c, \tag{9.5}$$

where $x(s) = \sum \mathscr{B}(s)$ is lexico minimal. □

Example 9.2 (Continued) Consider again the *chair* network of Fig. 8.2. Choose $q = 2$. For $s = 2$ we have $x(2) = \sum \mathscr{B}(2) = (2, 2, 1)$ and $Z(x(2), 2) = \frac{9}{25}$. For $s = 3$ we have $\tilde{x}(3) = \sum \tilde{\mathscr{B}}(3) = (3, 2)$ and $Z(\tilde{x}(3), 2) = \frac{13}{25}$. Over the range $s \in [2, 3]$, from Definition 9.3 we have $D_2 = \log(13/9)/\log(3/2) \approx 0.907$. For this network, not only is the value of D_q dependent on the minimal s-covering selected, but even the overall shape of the D_q vs. q curve depends on the minimal s-covering selection. For $x(2) = (2, 2, 1)$ we have

$$Z(x(2), q) = 2 \left(\frac{2}{5} \right)^q + \left(\frac{1}{5} \right)^q .$$

For $\tilde{x}(3) = (3, 2)$ we have

$$Z(\tilde{x}(3), q) = \left(\frac{3}{5} \right)^q + \left(\frac{2}{5} \right)^q .$$

Over the range $s \in [2, 3]$, from (9.5) we have

$$\tilde{D}_q \equiv \left(\frac{1}{q-1}\right) \left(\frac{\log\left(\frac{3^q + 2^q}{5^q}\right) - \log\left(\frac{(2)(2^q) + 1}{5^q}\right)}{\log(3/\Delta) - \log(2/\Delta)}\right) = \frac{\log\left(\frac{3^q + 2^q}{(2)(2^q) + 1}\right)}{\log(3/2)(q-1)} . \quad (9.6)$$

If for $s = 3$ we instead choose the covering $\hat{\mathscr{B}}(3)$ then for $\hat{x}(3) = (4, 1)$ we have

$$Z(\hat{x}(3), q) = \left(\frac{4}{5}\right)^q + \left(\frac{1}{5}\right)^q .$$

Again over the range $s \in [2, 3]$, but now using $\hat{x}(3)$ instead of $\tilde{x}(3)$, we obtain

$$\hat{D}_q \equiv \left(\frac{1}{q-1}\right) \left(\frac{\log\left(\frac{4^q + 1^q}{5^q}\right) - \log\left(\frac{(2)(2^q) + 1}{5^q}\right)}{\log(3/\Delta) - \log(2/\Delta)}\right) = \frac{\log\left(\frac{4^q + 1}{(2)(2^q) + 1}\right)}{\log(3/2)(q-1)} . \quad (9.7)$$

Figure 9.1 plots $\tilde{D}_q$ vs. q, and $\hat{D}_q$ vs. q over the range $0 \le q \le 15$. Neither curve is monotone non-increasing: the $\tilde{D}_q$ curve (corresponding to the lexico minimal summary vector $\tilde{x}(3) = (3, 2)$) is unimodal, with a local minimum at $q \approx 4.1$, and the $\hat{D}_q$ curve is monotone increasing.   □

The fact that neither curve in Fig. 9.1 is monotone non-increasing is remarkable, since it is well known that for a geometric multifractal, the D_q vs. q curve is monotone non-increasing [20]. The shape of the D_q vs. q curve will be explored further in Chap. 10. We next show that the $x(s)$ summary vectors can be used to compute $D_\infty \equiv \lim_{q \to \infty} D_q$. Let $x(s) = \sum \mathscr{B}(s)$ be lexico minimal, and let $x_1(s)$ be the first element of $x(s)$. It is proved in [48] that

$$\log\left(\frac{x_1(s)}{N}\right) \approx D_\infty \log\left(\frac{s}{\Delta}\right) . \quad (9.8)$$

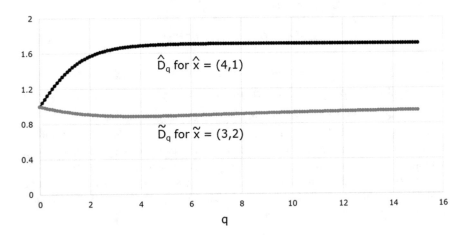

Fig. 9.1 Two plots of the generalized dimensions for the *chair* network

We can use (9.8) to compute D_∞ without having to compute any partition function values. It is well known [41] that, for geometric multifractals, D_∞ corresponds to the densest part of the fractal. Similarly, (9.8) shows that, for a complex network, D_∞ is the factor that relates the box size s to $x_1(s)$, the number of nodes in the box in the lexico minimal s-covering for which $p_j(s)$ is maximal.

To conclude this chapter, we consider the *sandbox method* for approximating D_q. The sandbox method, originally designed to compute D_q for geometric multifractals obtained by simulating diffusion-limited aggregation on a lattice [64, 65, 71], overcomes a well-recognized [1] limitation of using box counting to compute generalized dimensions: spurious results can be obtained for $q \ll 0$. This will happen if some box probability p_j is close to zero, for then when $q \ll 0$ the term p_j^q will dominate the partition sum $\sum_j p_j^q$. The sandbox method has also been shown to be more accurate than box counting for geometric fractals with known theoretical dimensions [62]. To describe the sandbox method, note that for a geometric multifractal for which D_q exists, by (9.1) and (9.2) we have, as $s \to 0$,

$$Z_q(\mathcal{B}(s)) = \sum_{B_j \in \mathcal{B}(s)} p_j^q(s) = \sum_{B_j \in \mathcal{B}(s)} p_j(s)[p_j(s)]^{q-1} \sim s^{(q-1)D_q} .$$

The sandbox method approximates $\sum_{B_j \in \mathcal{B}(s)} p_j^q(s)$ as follows [62]. Let $\widetilde{\mathbb{N}}$ be a randomly chosen subset of the N points and define $\widetilde{N} \equiv |\widetilde{\mathbb{N}}|$. With $M(n, r)$ defined by (7.1) and (7.2), define

$$avg(p^{q-1}(r)) \equiv \frac{1}{\widetilde{N}} \sum_{n \in \widetilde{\mathbb{N}}} \left(\frac{M(n, r)}{N}\right)^{q-1} , \tag{9.9}$$

where the notation $avg(p^{q-1}(r))$ is chosen to make it clear that this average uses equal weights of $1/\widetilde{N}$. Let L be the linear size of the lattice. The essence of the sandbox method is the approximation, for $r \ll L$,

$$avg(p^{q-1}(r)) \sim (r/L)^{(q-1)D_q} . \tag{9.10}$$

Note that $\sum_{B_j \in \mathcal{B}(s)} p_j^q(s)$ is a sum over the set of non-empty grid boxes, and the weight applied to $[p_j(s)]^{q-1}$ is $p_j(s)$. In contrast, $avg(p^{q-1}(r))$ is a sum over a randomly selected set of sandpiles, and the weight applied to $(M(n, r)/N)^{q-1}$ is $1/\widetilde{N}$. Since the $\widetilde{N}$ sandpile centers are chosen from the N points using a uniform distribution, the sandpiles may overlap. Because the sandpiles may overlap, and the sandpiles do not necessarily cover all the N points, in general $\sum_{n \in \widetilde{\mathbb{N}}} M(n, r) \neq N$, and we cannot regard the values $\{M(n, r)/N\}_{n \in \widetilde{\mathbb{N}}}$ as a probability distribution. Let β be the spacing between adjacent lattice positions (e.g., between adjacent horizontal and vertical positions for a lattice in $\mathbb{R}^2$).

Definition 9.4 For $q \neq 1$, the *sandbox dimension function* [62] of order q is the function of r defined for $\beta \leq r \ll L$ by

$$D_q^{sandbox}(r/L) \equiv \frac{1}{q-1} \frac{\log avg(p^{q-1}(r))}{\log(r/L)}. \tag{9.11}$$

$\square$

For a given $q \neq 1$ and lattice size L, the sandbox dimension function does not define a single sandbox dimension, but rather a range of sandbox dimensions, depending on r. It is not meaningful to define $\lim_{r \to 0} D_q^{sandbox}(r/L)$, since r cannot be smaller than the spacing β between lattice points. In practice, for a given q and L, a single value $D_q^{sandbox}$ of the sandbox dimension of order q is typically obtained by computing $D_q^{sandbox}(r/L)$ for a range of r values, and finding the slope of the $\log avg(p^{q-1}(r))$ vs. $\log(r/L)$ curve. The estimate of $D_q^{sandbox}$ is $1/(q-1)$ times this slope.

The sandbox method was applied to complex networks in [34]. The box centers are randomly selected nodes. There is no firm rule in [34] on the number of random centers to pick: they use $\tilde{N} \equiv |\tilde{\mathbb{N}}| = 1000$ random nodes, but suggest that $\tilde{N}$ can depend on N. For a given $q \neq 1$, they compute $avg(p^{q-1}(r))$ for a range of r values. Adapting (9.10) to a complex network $\mathbb{G}$, for $r \ll \Delta$ we have

$$\log avg(p^{q-1}(r)) \sim (q-1)D_q^{sandbox} \log(r/\Delta). \tag{9.12}$$

In [34], linear regression is applied to (9.12) to compute $D_q^{sandbox}$.

The sandbox method was applied to undirected weighted networks in [58]. The calculation of the sandbox radii in [58] is similar to the selection of box sizes discussed in Sect. 3.2.

Chapter 10
Non-monotonicity of Generalized Dimensions

In Chap. 9, we showed that the value of D_q for a given q depends in general on which minimal s-covering is selected, and we showed that this ambiguity can be eliminated by using the unique lexico minimal summary vectors $x(s)$. However, there remains a significant ambiguity in computing D_q, since Definition 9.3 refers to a range of s values over which approximate equality holds. Let this range be denoted by $[L, U]$, where $L < U$. It is well known that, in general, the numerical value of any fractal dimension depends on the range of box sizes over which the dimension is computed. What had not been previously recognized is that for a complex network the choice of L and U can dramatically change the shape of the D_q vs. q curve: depending on L and U, the shape of the D_q vs. q curve can be monotone increasing, or monotone decreasing, or even have both a local maximum and a local minimum [49]. Example 9.1 and Fig. 9.1 provided an example where the D_q vs. q plot is not monotone non-increasing, even for the simple case $[L, U] = [2, 3]$. This behavior stands in sharp contrast to the behavior of a geometric multifractal, for which it is known [20] that D_q is non-increasing in q.

Recalling that $\log Z(x(s), q)$ for a complex network $\mathbb{G}$ is defined by (9.4), one way to compute D_q for a given q is to determine a range $[L_q, U_q]$ of s over which $\log Z(x(s), q)$ is approximately linear in $\log s$, and then use (9.5) to estimate D_q, e.g., using linear regression. With this approach, to report computational results to other researchers, it would be necessary to specify, for each q, the range of box sizes used to estimate D_q. This is certainly not the protocol currently followed in research on generalized dimensions. Rather, the approach taken in [49] and [67] is to pick a single L and U and estimate D_q for all q with this L and U. Moreover, rather than estimating D_q using a technique such as regression over the range $[L, U]$ of box sizes, [49] instead estimates D_q using only the two box sizes L and U. (As discussed in Chap. 5, such a two-point estimate was also used in [46], where it was shown that even for as simple a network as a one-dimensional chain, estimates of d_C obtained from regression do not behave well, and a two-point estimate has very desirable properties.)

© The Author(s), under exclusive licence to Springer International Publishing AG, part of Springer Nature 2018
E. Rosenberg, *A Survey of Fractal Dimensions of Networks*, SpringerBriefs in Computer Science, https://doi.org/10.1007/978-3-319-90047-6_10

With this two-point approach, the estimate of D_q is $1/(q-1)$ times the slope of the secant line connecting the points

$$\big(\log L, \log Z(x(L),q)\big) \quad \text{and} \quad \big(\log U, \log Z(x(U),q)\big),$$

where $x(L)$ and $x(U)$ are the lexico minimal summary vectors for box sizes L and U, respectively. Using (9.4) and (9.5), this secant estimate of D_q, which we denote by $D_q(L, U)$, is defined by

$$D_q(L, U) \equiv \frac{\log Z(x(U),q) - \log Z(x(L),q)}{(q-1)\big(\log(U/\varDelta) - \log(L/\varDelta)\big)}$$

$$= \frac{1}{(q-1)\log(U/L)} \log\left(\frac{\sum_{B_j \in \mathscr{B}(U)}[x_j(U)]^q}{\sum_{B_j \in \mathscr{B}(L)}[x_j(L)]^q}\right). \qquad (10.1)$$

Example 10.1 Figure 10.1 plots box counting results for the *dolphins* network, which has 62 nodes, 159 arcs, and $\varDelta = 8$. This is a social network describing frequent associations between 62 dolphins in a community living off Doubtful Sound, New Zealand [35]. For this network, and for all other networks described in this chapter, each lexico minimal summary vector $x(s)$ was computed using Procedure 9.1 and the graph coloring heuristic described in [48]. Figure 10.1 shows that the $\big(-\log(s/\varDelta), \log B(s)\big)$ curve is approximately linear for $2 \le s \le 6$.

Figure 10.2 plots $\log Z(x(s),q)$ vs. $\log(s/\varDelta)$ for $2 \le s \le 6$ and for $q = 2, 4, 6, 8, 10$ ($q = 2$ is the top curve, and $q = 10$ is the bottom curve). Figure 10.2 shows that, although the $\log Z(x(s),q)$ vs. $\log(s/\varDelta)$ curves are less linear as q increases, a linear approximation is quite reasonable. Moreover, we are particularly interested in the behavior of the $\log Z(x(s),q)$ vs. $\log(s/\varDelta)$ curve for small positive q, the region where the linear approximation is best. Using (10.1), Fig. 10.3 plots the secant estimate $D_q(L, U)$ vs. q for various choices of L and U. Since the D_q vs. q

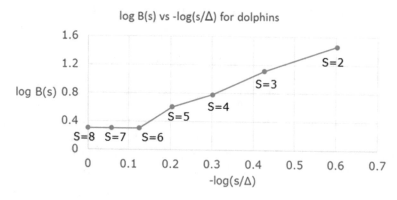

Fig. 10.1 Box counting for the *dolphins* network

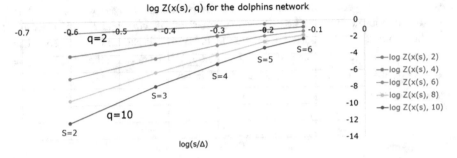

Fig. 10.2 $\log Z\big(x(s), q\big)$ vs. $\log(s/\Delta)$ for the *dolphins* network

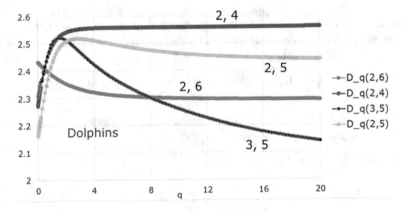

Fig. 10.3 Secant estimate of D_q for the *dolphins* network for different (L, U)

curve for a geometric multifractal is monotone non-increasing, it is remarkable that different choices of L and U lead to such different shapes for the $D_q(L, U)$ vs. q curve for the *dolphins* network.   □

Let $D_0'(L, U)$ denote the first derivative with respect to q of the secant $D_q(L, U)$, evaluated at $q = 0$. A simple closed-form expression for $D_0'(L, U)$ is derived in [49]. For box size s, let $x(s) = \sum \mathscr{B}(s)$ be lexico minimal. Define

$$G(s) \equiv \left(\prod_{j=1}^{B(s)} x_j(s) \right)^{1/B(s)}$$

$$A(s) \equiv \frac{1}{B(s)} \sum_{j=1}^{B(s)} x_j(s)$$

$$R(s) \equiv \frac{G(s)}{A(s)} \tag{10.2}$$

so $G(s)$ is the geometric mean of the box masses summarized by $x(s)$, $A(s)$ is the arithmetic mean of the box masses summarized by $x(s)$, and $R(s)$ is the ratio of the geometric mean to the arithmetic mean. By the classic arithmetic-geometric inequality, for each s we have $R(s) \leq 1$. Since $\sum_{j=1}^{B(s)} x_j(s) = N$, then $B(s) A(s) = N$. Theorems 10.1 and 10.2 below are proved in [49].

Theorem 10.1

$$D_0'(L, U) = \frac{1}{\log(U/L)} \log \frac{R(L)}{R(U)}.$$

$\square$

Theorem 10.1 says that the slope of the secant estimate of D_q at $q = 0$ depends on $x(L)$ and $x(U)$ only through the ratio of the geometric mean to the arithmetic mean of the components of $x(L)$, and similarly for $x(U)$. Since $L < U$, Theorem 10.1 immediately implies the following corollary.

Corollary 10.1 $D_0'(L, U) > 0$ *if and only if* $R(L) > R(U)$, *and* $D_0'(L, U) < 0$ *if and only if* $R(L) < R(U)$. $\square$

For a given L and U, Theorem 10.2 below provides a sufficient condition for $D_q(L, U)$ to have a local maximum or minimum.

Theorem 10.2 (i) *If* $R(L) > R(U)$ *and*

$$\frac{B(L)}{B(U)} > \frac{x_1(U)}{x_1(L)}$$

then $D_q(L, U)$ *has a local maximum at some* $q > 0$. (ii) *If* $R(L) < R(U)$ *and*

$$\frac{B(L)}{B(U)} < \frac{x_1(U)}{x_1(L)}$$

then $D_q(L, U)$ *has a local minimum at some* $q > 0$. $\square$

Example 10.2 To illustrate Theorem 10.2, consider the *dolphins* network of Example 10.1 with $L = 3$ and $U = 5$. We have $B(3) = 13$ and $B(5) = 4$, so $D_0 = \log(13/4)/\log(5/3) \approx 2.307$. Also, $x_1(3) = 10$ and $x_1(5) = 28$, so by (9.8) we have $D_\infty \approx \log(28/10)/\log(5/3) \approx 2.106$. We have $R(3) \approx 0.773$, $R(5) \approx 0.660$, and $D_0'(L, U) \approx 0.311$. Hence $D_q(3, 5)$ has a local maximum, as seen in Fig. 10.3. Moreover, for the *dolphins* network, choosing $L = 2$ and $U = 5$ we have $D_0 = \log(29/4)/\log(5/2) \approx 2.16$, and $D_\infty \approx \log(28/3)/\log(5/2) \approx 2.44$, so $D_0 < D_\infty$, as is evident from Fig. 10.3. Thus the inequality $D_0 \geq D_\infty$, which is valid for geometric multifractals, does not hold for the *dolphins* network with $L = 2$ and $U = 5$. $\square$

If for $s = L$ and $s = U$ we can compute a minimal s-covering with equal box masses, then $\mathbb{G}$ is a monofractal but not a multifractal. To see this, suppose all boxes in $\mathscr{B}(L)$ have the same mass, and that all boxes in $\mathscr{B}(U)$ have the same mass. Then for $s = L$ and $s = U$ we have $x_j(s) = N/B(s)$ for $1 \le j \le B(s)$, and (9.1) yields

$$Z(x(s), q) = \sum_{B_j \in \mathscr{B}(s)} \left(\frac{x_j(s)}{N} \right)^q = \sum_{B_j \in \mathscr{B}(s)} \left(\frac{1}{B(s)} \right)^q = [B(s)]^{1-q} .$$

From (9.5), for $q \ne 1$ we have

$$D_q = \frac{\log Z\big((x(U), q\big) - \log Z\big(x(L), q\big)}{(q - 1)\big(\log U - \log L\big)} = \frac{\log \big([B(U)]^{1-q}\big) - \log \big([B(L)]^{1-q}\big)}{(q - 1)\big(\log U - \log L\big)}$$

$$= \frac{\log B(L) - \log B(U)}{\log U - \log L} = D_0 = d_B , \tag{10.3}$$

so $\mathbb{G}$ is a monofractal. Thus equal box masses imply $\mathbb{G}$ is a monofractal, the simplest of all fractal structures.

There are several ways to try to obtain equal box masses in a minimal s-covering of $\mathbb{G}$. As discussed in Chap. 8, ambiguity in the choice of minimal coverings used to compute d_I is eliminated by maximizing entropy. Since the entropy of a probability distribution is maximized when all the probabilities are equal, a maximal entropy minimal covering equalizes (to the extent possible) the box masses. Similarly, as discussed in Chap. 9, ambiguity in the choice of minimal s-coverings used to compute D_q is eliminated by minimizing the partition function $Z_q(\mathscr{B}(s))$. Since for all sufficiently large q the lexico minimal vector $x(s)$ summarizes the s-covering that minimizes $Z_q(\mathscr{B}(s))$, and since for $q > 1$ a partition function is minimized when all the probabilities are equal, then $x(s)$ also equalizes (to the extent possible) the box masses. Theorem 10.1 suggests a third way to try to equalize the masses of all boxes in a minimal s-covering: since $G(s) \le A(s)$ and $G(s) = A(s)$ when all boxes have the same mass, a minimal s-covering that maximizes $G(s)$ will also equalize (to the extent possible) the box masses. The advantage of computing the lexico minimal summary vectors $x(s)$, rather than maximizing the entropy or maximizing $G(s)$, is that, by Theorem 9.1, the summary vector $x(s)$ is unique.

We now apply Theorem 10.1 to the *chair* network, to the *dolphins* network, and to a *jazz* network.

Example 10.3 For the *chair* network of Fig. 8.2 we have $L = 2$, $x(L) = (2, 2, 1)$, $U = 3$, and $x(U) = (3, 2)$. We have $D_0'(2, 3) \approx -0.070$, as shown in Fig. 9.1 by the slightly negative slope of the lower curve at $q = 0$. As mentioned above, this curve is not monotone non-increasing; it has a local minimum.   □

Example 10.4 For the *dolphins* network studied in Example 10.1, Table 10.1 provides $D_0'(L, U)$ for various choices of L and U. The values in Table 10.1 are

Table 10.1 $D'_0(L, U)$ for the *dolphins* network

L, U	$D'_0(L, U)$
2, 6	−0.056
3, 5	0.311
2, 4	0.393
2, 5	0.367

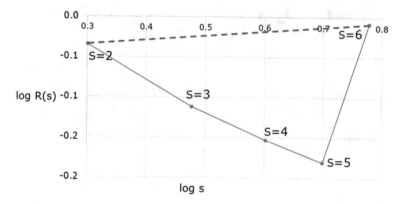

Fig. 10.4 log $R(s)$ vs. log s for the *dolphins* network

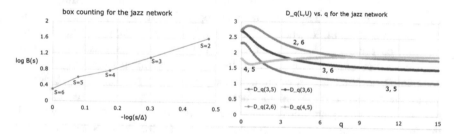

Fig. 10.5 *Jazz* box counting (left) and D_q vs. q for various L and U (right)

better understood using Fig. 10.4, which plots log $R(s)$ vs. log s. For example, for $(L, U) = (2, 6)$ we have $D'_0(2, 6) = \log\big(R(2)/R(6)\big)/\big(\log 6/2\big) \approx -0.056$, as illustrated by the slightly *positive* slope of the dashed red line in Fig. 10.4, since the slope of the dashed red line is $-D'_0(2, 6)$. For the other choices of (L, U) in Table 10.1, the values of $D'_0(L, U)$ are positive and roughly equal. Figure 10.2 visually suggests that log $Z\big(x(s), q\big)$ is better approximated by a linear fit over $s \in [2, 5]$ than over $s \in [2, 6]$, and Fig. 10.4 clearly shows that $s = 6$ is an outlier in that using $U = 6$ dramatically changes $D'_0(L, U)$. □

Example 10.5 This network, with 198 nodes, 2742 arcs, and diameter 6, is a collaboration network of jazz musicians [19]. Figure 10.5 shows the results of box counting; the curve appears reasonably linear for $s \in [2, 6]$. Figure 10.5 also plots $D_q(L, U)$ vs. q for four choices of L and U. Table 10.2 provides $D'_0(L, U)$, D_0, and D_∞ for nine choices of L and U; the rows are sorted by decreasing $D'_0(L, U)$.

Table 10.2 Results for the *jazz* network for various L and U

L, U	$D_0'(L, U)$	D_0	D_∞
2, 3	1.576	2.77	2.16
2, 4	1.224	2.74	1.42
2, 5	0.826	2.51	1.51
3, 4	0.728	2.69	0.37
2, 6	0.485	2.73	1.68
3, 5	0.231	2.31	1.00
3, 6	−0.154	2.70	1.40
4, 5	−0.411	1.82	1.82
5, 6	−1.232	3.80	2.52

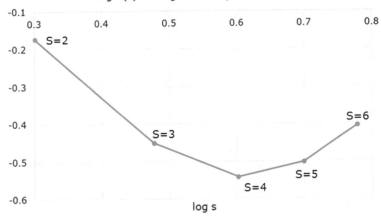

Fig. 10.6 $\log R(s)$ vs. $\log s$ for the *jazz* network

It is even possible for the $D_q(L, U)$ vs. q curve to exhibit both a local maximum and a local minimum: for the *jazz* network with $L = 4$ and $U = 5$, there is a local minimum at $q \approx 0.7$ and a local maximum at $q \approx 12.8$. Figure 10.6 plots $\log R(s)$ vs. $\log s$ for the *jazz* network. □

   These results, together with the results in [47, 48], show that two requirements should be met when reporting fractal dimensions of a complex network. First, since there are in general multiple minimal s-coverings, and these different coverings can yield different values of D_q, computational results should specify the rule (e.g., a maximal entropy covering, or a covering yielding a lexico minimal summary vector) used to unambiguously select a minimal s-covering. Second, the lower bound L and upper bound U on the box sizes used to compute D_q should be reported. Published values of D_q not meeting these two requirements cannot in general be considered benchmarks. As to the values of L and U yielding the most meaningful results, it is desirable to identify the largest range $[L, U]$ over which $\log Z$ is approximately linear in $\log s$; this is a well-known principle in the estimation of fractal dimensions. Future research may uncover, based on the $\log R(s)$ vs. $\log s$ curve, other criteria for selecting L and U.

Chapter 11
Zeta Dimension

In this final chapter we consider the use of the zeta function

$$\zeta(\alpha) = \sum_{i=1}^{\infty} i^{-\alpha} \qquad (11.1)$$

to define the dimension of a network. The zeta function has a rich history [8, 17]. It was studied by Euler in 1737 for non-negative real α and extended in 1859 by Riemann to complex α. We will consider the zeta function only for non-negative real α. The zeta function converges for $\alpha > 1$ and diverges otherwise. It is a decreasing function of α and $\zeta(\alpha) \to 1$ as $\alpha \to \infty$ [54].

The zeta function has been used to define the fractal dimension of a finite complex network [54]. Although the zeta dimension of a network has not enjoyed widespread popularity, it has interesting connections to the Hausdorff dimension. Recall from (7.5) in Chap. 7 that $\partial \mathbb{N}(n, r)$ is the set of nodes whose distance from node n is exactly r, and $|\partial \mathbb{N}(n, r)|$ is the number of such nodes. Define the *graph surface function* by

$$S_r \equiv \frac{1}{N} \sum_{n \in \mathbb{N}} |\partial \mathbb{N}(n, r)|, \qquad (11.2)$$

so S_r is the average number of nodes at a distance r from a random node in the network. Define the *graph zeta function* $\zeta_{\mathbb{G}}(\alpha)$ [54] by

$$\zeta_{\mathbb{G}}(\alpha) \equiv \frac{1}{N} \sum_{x \in \mathbb{N}} \sum_{\substack{y \in \mathbb{N} \\ y \neq x}} dist(x, y)^{-\alpha}. \qquad (11.3)$$

© The Author(s), under exclusive licence to Springer International Publishing AG, part of Springer Nature 2018
E. Rosenberg, *A Survey of Fractal Dimensions of Networks*, SpringerBriefs in Computer Science, https://doi.org/10.1007/978-3-319-90047-6_11

Since $\mathbb{G}$ is a finite network, then $\zeta_{\mathbb{G}}(\alpha)$ is finite, and the graph zeta function and the graph surface function are related by

$$\zeta_{\mathbb{G}}(\alpha) = \frac{1}{N} \sum_{x \in \mathbb{N}} \sum_{r \geq 1} |\partial \mathbb{N}(x, r)| r^{-\alpha}$$

$$= \sum_{r \geq 1} \left(\frac{1}{N} \sum_{x \in \mathbb{N}} |\partial \mathbb{N}(x, r)| \right) r^{-\alpha}$$

$$= \sum_{r \geq 1} S_r r^{-\alpha} . \qquad (11.4)$$

The function $\zeta_{\mathbb{G}}(\alpha)$ is decreasing in α. For $\alpha = 0$ we have

$$\zeta_{\mathbb{G}}(0) = (1/N) \sum_{x \in \mathbb{N}} (N - 1) = N - 1 .$$

For a given $x \in \mathbb{N}$, if $dist(x, y) > 1$ then $dist(x, y)^{-\alpha} \to 0$ as $\alpha \to \infty$, so

$$\lim_{\alpha \to \infty} \sum_{\substack{y \in \mathbb{N} \\ y \neq x}} dist(x, y)^{-\alpha} = \lim_{\alpha \to \infty} \sum_{\substack{y \in \mathbb{N} \\ dist(x,y)=1}} dist(x, y)^{-\alpha} = \delta_x , \qquad (11.5)$$

where δ_x is the node degree of x. Thus $\zeta_{\mathbb{G}}(\alpha)$ approaches the average node degree as $\alpha \to \infty$.

Since (11.3) defines $\zeta_{\mathbb{G}}(\alpha)$ only for a finite network, we would like to define $\zeta_{\mathbb{G}}(\alpha)$ for an infinite network $\mathbb{G}$. If $\mathbb{G} = \lim_{N \to \infty} \mathbb{G}_N$, where $\mathbb{G}_N$ has N^E nodes, then we can define $\zeta_{\mathbb{G}}(\alpha) \equiv \lim_{N \to \infty} \zeta_{\mathbb{G}_N}(\alpha)$, as is implicitly done in [54]. For example, $\mathbb{G} = \lim_{N \to \infty} \mathbb{G}_N$ holds when $\mathbb{G}$ is an infinite E-dimensional rectilinear lattice and $\mathbb{G}_N$ is a finite E-dimensional rectilinear lattice for which each edge has N nodes. Table 11.1, from [54], provides $\zeta_{\mathbb{G}}(\alpha)$ for an infinite rectilinear lattice in $\mathbb{R}^E$ and the L_1 norm. Here Γ denotes the gamma function, so $\Gamma(E) = (E - 1)!$.

We are interested in infinite networks $\mathbb{G}$ for which $\zeta_{\mathbb{G}}(\alpha)$ can be infinite. Since $\zeta_{\mathbb{G}}(\alpha)$ is a decreasing function of α, if $\zeta_{\mathbb{G}}(\alpha)$ is finite for some value α, it is finite for $\alpha' > \alpha$. If $\zeta_{\mathbb{G}}(\alpha)$ is infinite for some value α, it is infinite for $\alpha' < \alpha$. Thus there is at

Table 11.1 S_r and $\zeta_{\mathbb{G}}(\alpha)$ for an infinite rectilinear lattice in $\mathbb{R}^E$

E	S_r	$\zeta_{\mathbb{G}}(\alpha)$
1	2	$2\zeta(\alpha)$
2	$4r$	$4\zeta(\alpha - 1)$
3	$4r^2 + 2$	$4\zeta(\alpha - 2) + 2\zeta(\alpha)$
4	$(8/3)r^3 + (16/3)r$	$(8/3)\zeta(\alpha - 3) + (16/3)\zeta(\alpha - 1)$
$r \to \infty$	$O\left(2^E r^{E-1}/\Gamma(E)\right)$	$O\left(2^E \zeta(\alpha - E + 1)/\Gamma(E)\right)$

most one value of α for which $\zeta_{\mathbb{G}}(\alpha)$ transitions from being infinite to finite, and the *zeta dimension* d_Z of $\mathbb{G}$ is the value at which this transition occurs. This definition parallels the definition in Sect. 1.2 of the Hausdorff dimension as that value of d for which $v^*(d)$ transitions from infinite to finite. If for all α we have $\zeta_{\mathbb{G}}(\alpha) = \infty$ then d_Z is defined to be ∞.

Example 11.1 Let $\mathbb{G}$ be an infinite rectilinear lattice in $\mathbb{R}^3$. From Table 11.1 we have

$$\zeta_{\mathbb{G}}(\alpha) = 4\zeta(\alpha - 2) + 2\zeta(\alpha).$$

Since $4\zeta(\alpha - 2) + 2\zeta(\alpha) < \infty$ only for $\alpha > 3$, then

$$d_Z = \inf\{\alpha \mid \zeta_{\mathbb{G}}(\alpha) < \infty\} = 3. \quad \square$$

Example 11.2 As in [55], consider a random graph in which each pair of nodes is connected with probability p. For any three distinct nodes x, y, and z, the probability that z is not connected to both x and y is $1 - p^2$. The probability that x and y are not both connected to some other node is $(1 - p^2)^{N-2}$, which approaches 0 as $N \to \infty$. Thus for large N each pair of nodes is almost surely connected by a path of length at most 2. For large N, each node has $p(N - 1)$ neighbors, so from (11.2) we have $S_1 \approx p(N - 1)$. For large N, the number S_2 of nodes at distance 2 from a random node is given by $S_2 \approx (N - 1) - S_1 = (N - 1)(1 - p)$. Hence

$$\zeta_{\mathbb{G}_N}(\alpha) \approx p(N - 1) + (N - 1)(1 - p)2^{-\alpha}.$$

Since $\lim_{N \to \infty} \zeta_{\mathbb{G}_N}(\alpha) = \infty$ for all α then $d_Z = \infty$. $\quad \square$

An alternative definition of the dimension of an infinite graph, using the zeta function, but not requiring averages over all the nodes of the graph, is given in [55]. For $n \in \mathbb{N}$, define

$$\zeta_{\mathbb{G}}(n, \alpha) = \sum_{\substack{x \in \mathbb{N} \\ x \neq n}} dist(n, x)^{-\alpha}.$$

There is exactly one value of α at which $\zeta_{\mathbb{G}}(n, \alpha)$ transitions from being infinite to finite; denote this value by $d_Z(n)$. The alternative definition of the zeta dimension is

$$d_Z \equiv \limsup_{n \in \mathbb{N}} d_Z(n).$$

This definition is not always identical to the above definition of d_Z as the value at which $\zeta_{\mathbb{G}}(\alpha)$ transitions infinite to finite [55].

References

1. Ariza-Villaverde, A.B., Jiménez-Hornero, F.J., and De Ravé, E.G. (2013). Multifractal Analysis of Axial Maps Applied to the Study of Urban Morphology. *Computers, Environment and Urban Systems*, 38: 1–10.
2. Bais, F.A. and Farmer, J.D. (2008). The Physics of Information. Chapter in *Philosophy of Information*, edited by Adriaans P. and van Benthem J. (Elsevier, Oxford, UK).
3. Berntson, G.M. and Stoll, P. (1997). Correcting for Finite Spatial Scales of Self-Similarity when Calculating the Fractal Dimensions of Real-World Structures. *Proc. Royal Society London B*, 264: 1531–1537.
4. Broido, A.D. and Clauset, A. (2018). Scale-Free Networks are Rare. *arXiv:1801.03400 [physics.soc-ph]*
5. Chvátal, V. (1979). A Greedy Heuristic for the Set Covering Problem. *Mathematics of Operations Research*, 4: 233–235.
6. Chvátal, V. (1983) *Linear Programming* (W.H. Freeman, New York).
7. da F. Costa, L., Rodrigues, F. A., Travieso, G. and Villas Boas, P.R. (2007). Characterization of Complex Networks: A Survey of Measurements. *Advances in Physics*, 56: 167–242.
8. Doty, D., Gu, X., Lutz, J.H., Mayordomo, E., and Moser, P. (2005). Zeta-Dimension. Chapter in *Mathematical Foundations of Computer Science 2005* (Springer, New York).
9. Eckmann, J.P. and Ruelle, D. (1985). Ergodic Theory of Chaos and Strange Attractors. *Reviews of Modern Physics*, 57: 617–656.
10. Eguiluz, V.M., Hernandez-Garcia, E., Piro, O., and Klemm, K. (2003). Effective Dimensions and Percolation in Hierarchically Structured Scale-Free Networks. *Physical Review E*, 68: 055102(R).
11. Erlenkotter, D. (1978). A Dual-Based Procedure for Uncapacitated Facility Location. *Operations Research*, 26: 992–1009 (errata in *Operations Research*, 28 (1980) p. 442).
12. Falconer, K. (2003). *Fractal Geometry: Mathematical Foundations and Applications*, 2nd edn. (Wiley, West Sussex, England).
13. Farmer, J.D. (1982). Information Dimension and the Probabilistic Structure of Chaos. *Z. Naturforsch.*, 37a: 1304–1325.
14. Farmer, J.D., Ott, E., and Yorke, J.A. (1983). The Dimension of Chaotic Attractors. *Physica*, 7D: 153–180.
15. Furuya, S. and Yakubo, K. (2011). Multifractality of Complex Networks. *Physical Review E*, 84: 036118.
16. Gallos, L.K., Song, C., and Makse, H.A. (2007). A Review of Fractality and Self-Similarity in Complex Networks. *Physica A*, 386: 686–691.

17. Garrido, A. (2009). Combinatorial Analysis by the Ihara Zeta Function of Graphs. *Advanced Modeling and Optimization*, 11: 253–278.

18. Gill, P.E., Murray, W., and Wright, M.H. (1981). *Practical Optimization* (Academic Press, Bingley, UK).

19. Gleiser, P.M. and Danon, L. (2003). Community Structure in Jazz. *Advances in Complex Systems*, 6: 565. Data available at http://konect.uni-koblenz.de/networks/arenas-jazz.

20. Grassberger, P. (1983). Generalized Dimensions of Strange Attractors. *Physics Letters*, 97A: 227–230.

21. Grassberger, P. (1985). Generalizations of the Hausdorff Dimension of Fractal Measures. *Physics Letters A*, 107: 101–105.

22. Grassberger P. and Procaccia, I. (1983). Characterization of Strange Attractors. *Physical Review Letters*, 50: 346–349.

23. Grassberger, P. and Procaccia, I. (1983). Measuring the Strangeness of Strange Attractors. *Physica*, 9D: 189–208.

24. Hausdorff, F. (1919). Dimension and ausseres Mass. *Math. Annalen*, 79: 157–179.

25. Hentschel, H.G.E. and Procaccia, I. (1983). The Infinite Number of Generalized Dimensions of Fractals and Strange Attractors. *Physica D*, 8: 435–444.

26. Jaynes, E.T. (1957). Information Theory and Statistical Mechanics. *The Physical Review*, 106: 620–630.

27. Jelinek, H.F. and Fernandez, E. (1998). Neurons and Fractals: How Reliable and Useful are Calculations of Fractal Dimensions? *Journal of Neuroscience Methods*, 81: 9–18.

28. Kenkel, N.C. (2013). Sample Size Requirements for Fractal Dimension Estimation. *Community Ecology*, 14: 144–152.

29. Kim, J.S., Goh, K.I., Kahng, B., and Kim, D. (2007). A Box-Covering Algorithm for Fractal Scaling in Scale-Free Networks. *Chaos*, 17: 026116.

30. Kim, J.S., Goh, K.I., Kahng, B., and Kim, D. (2007). Fractality and Self-Similarity in Scale-Free Networks. *New Journal of Physics*, 9: 177.

31. Klarreich, E. (2018). Scant Evidence of Power Laws Found in Real-World Networks. *Quanta Magazine*, February 15, 2018.

32. Lacasa, L. and Gómez-Gardeñes, J. (2013). Correlation Dimension of Complex Networks. *Physical Review Letters*, 110: 168703.

33. Lacasa, L. and Gómez-Gardeñes, J. (2014). Analytical Estimation of the Correlation Dimension of Integer Lattices. *Chaos*, 24: 043101.

34. Liu, J.L., Yu, Z.G., and Anh, V. (2015). Determination of Multifractal Dimensions of Complex Networks by Means of the Sandbox Algorithm. *Chaos*, 25: 023103.

35. Lusseau,D., Schneider, K., Boisseau, O. J., Haase, P., Slooten, E., and Dawson, S. M. (2003). The Bottlenose Dolphin Community of Doubtful Sound Features a Large Proportion of Long-Lasting Associations. *Behavioral Ecology and Sociobiology*, 54: 396–405.

36. Mandelbrot, B.B. (1983). *The Fractal Geometry of Nature* (W.H. Freeman, New York).

37. Newman, M.E.J. Network Data. http://www-personal.umich.edu/~mejn/netdata/.

38. Nowotny, T. and Requardt, M. (1988). Dimension Theory of Graphs and Networks. *J. Phys. A: Math. Gen.*, 31: 2447–2463.

39. Orozco, C.D.V., Golay, J., and Kanevski, M. (2015). Multifractal Portrayal of the Swiss Population. *Cybergeo: European Journal of Geography*, 714 http://cybergeo.revues.org/26829.

40. Paladin, G. and Vulpiani, A. (1987). Anomalous Scaling Laws in Multifractal Objects. *Physics Reports*, 156: 147–225.

41. Peitgen, H.O., Jürgens, H., and Saupe, D. (1992) *Chaos and Fractals* (Springer-Verlag, New York).

42. Rosenberg, E. (2001). Dual Ascent for Uncapacitated Telecommunications Network Design with Access, Backbone, and Switch Costs. *Telecommunications Systems*, 16: 423–435.

43. Rosenberg, E. (2012). *A Primer of Multicast Routing* (Springer, New York).

44. Rosenberg, E. (2013). Lower Bounds on Box Counting for Complex Networks. *Journal of Interconnection Networks*, 14: 1350019.

45. Rosenberg, E. (2016). Minimal Box Size for Fractal Dimension Estimation. *Community Ecology*, 17: 24–27.
46. Rosenberg, E. (2016). The Correlation Dimension of a Rectilinear Grid. *Journal of Interconnection Networks*, 16: 1550010.
47. Rosenberg, E. (2017). Maximal Entropy Coverings and the Information Dimension of a Complex Network. *Physics Letters A*, 381: 574–580.
48. Rosenberg, E. (2017). Minimal Partition Coverings and Generalized Dimensions of a Complex Network. *Physics Letters A*, 381: 1659–1664.
49. Rosenberg, E. (2017). Non-monotonicity of the Generalized Dimensions of a Complex Network. *Physics Letters A*, 381: 2222–2229.
50. Rosenberg, E. (2017). Erroneous Definition of the Information Dimension in Two Medical Applications. *International Journal of Clinical Medicine Research*, 4: 72–75.
51. Rozenfeld, H.D., Gallos, L. K., Song, C., and Makse, H.A. (2009) Fractal and Transfractal Scale-Free Networks. Chapter in *Encyclopedia of Complexity and Systems Science*, edited by R.A. Meyers, (Springer-Verlag, New York): 3924–3943.
52. Ruelle, D. (1990). Deterministic Chaos: The Science and the Fiction (The 1989 Claude Bernard Lecture). *Proc. R. Soc. Lond. A*, 427: 241–248.
53. Schleicher, D. (2007). Hausdorff Dimension, Its Properties, and Its Surprises. *The American Mathematical Monthly*, 114: 509–528.
54. Shanker, O. (2007). Graph Zeta Function and Dimension of Complex Network. *Modern Physics Letters B*, 21: 639–644.
55. Shanker, O. (2013). Dimension Measure for Complex Networks. Chapter in *Advances in Network Complexity*, edited by M. Dehmer, A. Mowshowitz, and F. Emmert-Streib (Wiley-VCH Verlag GmbH & Co. KGaA).
56. Song, C., Gallos, L.K., Havlin, S., and Makse, H.A. (2007). How to Calculate the Fractal Dimension of a Complex Network: the Box Covering Algorithm. *Journal of Statistical Mechanics*: P03006.
57. Song, C., Havlin, S., and Makse, H.A. (2005). Self-similarity of Complex Networks. *Nature*, 433: 392–395.
58. Song, Y.Q., Liu, J.L., Yu, Z.G., and Li, B.G. (2015). Multifractal Analysis of Weighted Networks by a Modified Sandbox Algorithm. *Scientific Reports*, 5: 17628.
59. Stanley H.E. and Meakin, P. (1998). Multifractal Phenomena in Physics and Chemistry. *Nature*, 335: 405–409.
60. Sun, Y. and Zhao, Y. (2014). 'Overlapping-box-covering Method for the Fractal Dimension of Complex Networks. *Physical Review E*, 89: 042809.
61. Tél, T. (1988). Fractals, Multifractals, and Thermodynamics: An Introductory Review. *Z. Naturforsch*, 43a: 1154–1174.
62. Tél, T., Fülöp, Á., and Vicsek, T. (1989). Determination of Fractal Dimensions for Geometrical Multifractals. *Physica A*, 159: 155–166.
63. Theiler, J., (1990). Estimating Fractal Dimension. *J. Optical Society of America A*, 7: 1055–1073.
64. Vicsek, T. (1989). *Fractal Growth Phenomena* (World Scientific, Singapore).
65. Vicsek T., Family, F., and Meakin, P. (1990). Multifractal Geometry of Diffusion Limited Aggregates. *Europhysics Letters*, 12: 217–222.
66. Wang, X., Liu, Z., and Wang, M. (2013). The Correlation Fractal Dimension of Complex Networks. *International Journal of Modern Physics C*, 24: 1350033.
67. Wang, D.L., Yu, Z.G., and Anh, V. (2011) Multifractality in Complex Networks. *Chinese Physics B*, 21: 080504.
68. Wei, D., Chen, X., and Deng, Y. (2016). Multifractality of Weighted Complex Networks. *Chinese Journal of Physics*, 54: 416–423.
69. Wei, D.J., Liu, Q., Zhang, H.X., Hu, Y., Deng, Y. and Mahadevan, S. (2013). Box-Covering Algorithm for Fractal Dimension of Weighted Networks. *Scientific Reports*, 3: https://doi.org/10.1038/srep03049.

70. Wei, D., Wei, B., Hu, Y., Zhang, H., and Deng, Y. (2014). A New Information Dimension of Complex Networks. *Physics Letters A*, 378: 1091–1094.
71. Witten, T.A. and Sander, L.M. (1981). Diffusion-Limited Aggregation, a Kinetic Critical Phenomenon. *Physical Review Letters*, 47: 1400–1403.
72. Zhang, H. , Wei, D., Hu, Y., Lan, X., and Deng, Y. (2016). Modeling the Self-Similarity in Complex Networks based on Coulomb's Law. *Communications in Nonlinear Science and Numerical Simulation*, 35: 97–104.
73. Zhang, Z., Zhou, S., Chen, L., and Guan, J. (2008). Transition from Fractal to Non-Fractal Scalings in Growing Scale-Free Networks. *The European Physics Journal B*, 64: 277–283.
74. Zhou, W.X., Jiang, Z.Q., and Sornette, D. (2007). Exploring Self-Similarity of Complex Cellular Networks: The Edge-Covering Method with Simulated Annealing and Log-Periodic Sampling., *Physica A: Statistical Mechanics and its Applications*, 375: 741–752.